Ore Sampling For Small Scale Miners

by Bureau of Mines

with an introduction by Kerby Jackson

Introduction

It has been almost a century since the U.S. Bureau of Mines released their important publication "Conditions of Ore Sampling". First released in 1916, this work has been unavailable to the mining community since those days, with the exception of expensive original collector's copies and poorly produced digital editions.

It has often been said that *"gold is where you find it"*, but even beginning prospectors understand that their chances for finding something of value in the earth or in the streams of the Golden West are dramatically increased by going back to those places where gold and other minerals were once mined by our forerunners. Despite this, much of the contemporary information on local mining history that is currently available is mostly a result of mere local folklore and persistent rumors of major strikes, the details and facts of which, have long been distorted. Long gone are the old timers and with them, the days of first hand knowledge of the mines of the area and how they operated. Also long gone are most of their notes, their assay reports, their mine maps and personal scrapbooks, along with most of the surveys and reports that were performed for them by private and government geologists. Even published books such as this one are often retired to the local landfill or backyard burn pile by the descendents of those old timers and disappear at an alarming rate. Despite the fact that we live in the so-called "Information Age" where information is supposedly only the push of a button on a keyboard away, true insight into mining properties remains illusive and hard to come by, even to those of us who seek out this sort of information as if our lives depend upon it. Without this type of information readily available to the average independent miner, there is little hope that our metal mining industry will ever recover.

This important volume and others like it, are being presented in their entirety again, in the hope that the average prospector will no longer stumble through the overgrown hills and the tailing strewn creeks without being well informed enough to have a chance to succeed at his ventures.

Kerby Jackson
Josephine County, Oregon
January 2015

CONTENTS.

3

TABLES.

ILLUSTRATIONS.

ORE-SAMPLING CONDITIONS IN THE WEST.

By T. R. WOODBRIDGE.

INTRODUCTION.

The success of every mining enterprise depends on accurate knowledge of the constituents of the ore taken from the mine. This knowledge may be least necessary for a placer mine or for a company producing bullion from its own ores, but it is absolutely necessary to an individual or a company selling its ore, or to a smelter, custom mill, or sampling plant buying the ore.

In spite of the importance of sampling, a large part of the ore sampling of to-day is not done on a scientific basis. There are many operators seeking thoroughly satisfactory methods of sampling who, nevertheless, are not doing as accurate work as the present state of the industry permits. On the other hand, investigation shows that in the majority of sampling plants there is a great lack of uniformity, and that even in individual plants several methods or combinations of methods may be found.

Discussions concerning the theory and practice of ore sampling have appeared at various times in the publications of the engineering societies and the technical press, but have not resulted in any general uniformity of practice. Especially is this true in regard to discarding methods known to be unreliable. Doubtless some operators are satisfied with this state of affairs and are not concerned with the accuracy of the sampling of individual lots of ore, provided the total purchase price of all lots equals the total value recovered during a certain period of time. Also, a few operators have worked out the problem of sampling for their local conditions to their own satisfaction and profit, and therefore do not welcome any criticism that may result in a change of methods. Most sellers and buyers of ore, however, are frankly disgusted with the persistent use of incorrect methods and would welcome a thorough investigation and discussion of the subject.

In connection with its efforts to increase efficiency in the mineral industries, the Bureau of Mines has undertaken to investigate present methods of sampling and analyzing ores in this country and to present the facts to all who may be interested in their study and discussion.

The object of this report is to give the facts regarding the present methods of ore sampling in the hope that there may be discussion that will ultimately result in the standardization of sampling methods. Sampling would thus be raised from the position of an art to that of an exact science, where it properly belongs.

The first step forward would be the discarding of methods known to be inaccurate and of methods that, under certain conditions, may be manipulated to give inaccurate results to the advantage of either the seller or the buyer. This would eliminate one of the chief causes of the friction that often develops between the seller and the buyer. At any rate, each party to the transaction should be familiar with the limitations of any method that has been employed by mutual consent, so that neither party will feel that the other is obtaining an unfair advantage.

ACKNOWLEDGMENT.

Grateful acknowledgment is made to the officers and employees of the various plants visited for the information given, the suggestions offered, and the interest manifested in the writer's mission, and also for the many personal courtesies shown. The writer also wishes to acknowledge his indebtedness for many suggestions from former employers, from many employees, and from others with whom he came in contact during several years of active work in ore sampling and in the experimental work connected with the improvement of certain methods.

SCOPE OF THE INVESTIGATION.

In this report the subject of ore sampling is considered under five divisions, as follows: A definition of ore sampling; the condition of the ore as affecting sampling; description of the various methods of of sampling used in the West, with flow sheets showing their application, and a criticism of certain details of practice; discussion of the methods investigated; and recommendations. The report deals principally with the first three of these divisions, although it may occasionally encroach upon the other two. Under various headings it has seemed necessary to repeat or discuss more fully some of the conditions of ore sampling in order to emphasize their effects on certain sampling processes.

An exhaustive review or criticism is not attempted, and such deductions and suggestions as are made are offered with the knowledge that there must necessarily be some difference of opinion among well-informed and honest operators. It is therefore hoped that this paper will be freely discussed by those who can offer constructive criticism.

CORRECT ORE SAMPLING DEFINED.

The correct sampling of a lot of ore is the process of obtaining from it a smaller quantity that contains, in unchanged percentages, all the constituents of the original lot.

The commercial object of sampling is accomplished when the ultimate sample obtained meets the above conditions within an allowable limit of error, and has been obtained with reasonable speed and at a moderate cost. The final sample should be dry and of such bulk and degree of fineness as to be immediately available for the determination by the assayer or chemist of one or more of its constituents.

The commercial sample of an ore need not be more exact than the limits of error in the methods of determination employed by the assayer or chemist.

CONDITIONS AFFECTING ORE SAMPLING.

SEGREGATION IN SAMPLES.

The obtaining of a satisfactory sample of an ore would be a comparatively simple undertaking if the constituents to be determined were uniformly distributed throughout the whole mass. In most ores, however, the constituents, particularly the gold and silver, are more or less segregated. Ore in place in the mine is rarely of uniform metal content. The process of mining generally aggravates this condition, as the more heavily mineralized parts usually break up more readily, so that the finer particles frequently contain a higher percentage of the valuable constituents than the coarser particles.

Should the ore be screened before being sampled, the difference may be lessened but never entirely eliminated. With every crushing during the sampling process this difference will persist to a varying degree, even after the final pulp has passed through a 120-mesh screen. Occasionally the condition is reversed—that is, the gangue breaks up more readily than the more heavily mineralized ore—but this does not alter the problem that arises through the segregation of certain constituents of the ore during mining and sampling.

In either event, even after repeated crushing and rolling (crushing with rolls), there still exists an imperfect mixture of fine and coarse particles, ranging from impalpable dust to pieces several inches in diameter. Unfortunately no method for making an intimate and uniform mixture of ore particles of various sizes has ever been discovered. No amount of shoveling or coning and no form of revolving receptacle will accomplish more than to make the segregation somewhat less apparent to the eye. When such a mixture is taken up with a shovel, the coarser particles tend to roll off or to collect at the edges of the shovel; if thrown upon a cone, . the coarser particles tend to roll outward to the edge of the cone; if allowed to slide to the floor from an inclined chute, the coarser particles will run ahead and will accumulate farther from the edge of the chute than the finer, slower moving particles; if the ore falls . through a long spout, the larger particles will bound from side to side, while the smaller particles will proceed in a more direct course; if fed into a revolving barrel, the finer particles will be carried higher on the side of the barrel and will be discharged from a different point than the coarser particles; when the mixture is screened, the softer and finer particles pass through first.

In fact every time that the ore is handled the particles tend to segregate according to size. This tendency to segregate is frequently mentioned in this paper, as it forms the basis for most of the schemes for favoring either the buyer or seller of ore.

In considering the reliability of any sampling method this tendency to segregation should be constantly borne in mind. Any system or any detail of a system that may tend to make possible the selection or rejection of the finer or coarser particles should be considered inaccurate. So absolutely does the accuracy of a sampling method depend on freedom from selection that one may forget that gold, silver, and other metals are to be determined by the process and examine the method solely as to its ability to take from a lot of ore the same proportion of all the various sizes of particles.

LACK OF UNIFORMITY IN SAMPLING METHODS.

The success of the older systems of ore sampling is considered by many to depend on the possibility of mixing particles of ore varying widely in size and in the percentage of their various constituents, whereas in reality it depends on the uniformity with which these particles are finally distributed around the axis of a truncated cone.

With the further study of the subject and through the comparisons resulting from repeated sampling of the same product by the use of different methods, there have been developed the so-called "automatic" or "mechanical" systems of sampling, which are founded on the principle of taking small portions from the stream of ore at frequent intervals rather than large quantities at infrequent intervals. These systems may be said to be forms of fractional mixing of the particles of an ore, in that they aim to send to the final sample many thousands of easily mixed small portions, any one of which may not seriously affect the sample but which in the aggregate will approach the average value well within the limits of accuracy required.

A study or an examination of the sampling methods used in the Western States shows that, in a large proportion of the plants, the theory of sampling has been given proper consideration, and many sampling plants have been built and are being operated on correct principles. It also shows that there is great variation in the application of many methods and a general lack of uniformity in carrying out certain details, which often results in a ridiculous waste of time and money and causes ore sampling to drift far from its true scientific and technical position to one of either guesswork, a too great dependence on the law of long period averages, or, in certain instances, to a plain case of matching wits.

These conditions are indicated by the fact that a method considered reliable in one plant may be considered unreliable in another, although the two plants may belong to one company; that much resampling and reassaying is considered necessary before an agreement is reached between the buyer and the seller; and that sampling often provides merely a means of settlement rather than a correct estimate of the contents of the ore. Therefore to-day, as in many years past, there are the buyer who by accident or intent impoverishes the sample with too large a proportion of the lower grade, coarser particles, and the seller who screens and remixes his ores in order that the buyer may cheat himself.

MULTISAMPLING.

In many of the sampling plants that handle the more valuable ores are men who, although they refer with satisfaction to certain marvelously close checks in some resamplings, nevertheless take for granted that a large proportion of the lots received will necessarily and unavoidably show differences that will necessitate a second or third resampling. It is the custom in many plants to divide the sample at certain points in the process, and thus make what are known as duplicate, triplicate, and even quadruplicate samples, each of which receives an entirely independent but similar treatment during the later processes and is assayed separately. This is done for several reasons, one being that it furnishes a means for checking any gross error in the succeeding operations, and another being the possibility of using an average of the results in making a settlement. In some agreements between buyer and seller it is provided that the difference between these multisamples be used as a guide in determining whether resampling of the ore is necessary. The necessity for resampling is sometimes decided on the merits of each particular case, or, in some localities, a certain agreed percentage difference or tolerance between an original and a duplicate is allowed before a resample may be demanded. In one instance a 20 per cent difference is considered a satisfactory check, but in the majority of such agreements a 10 per cent difference automatically calls for a resample, whereas any difference less than 10 per cent precludes resampling and forces a settlement on the basis of the original or an umpire assay. For example, a buyer working under a 10 per cent agreement is satisfied as to the correctness of the sampling of an ore containing 5 ounces of gold per ton if one sample, the original, assays 5 ounces and the other, the duplicate, assays either 4.54 or 5.50 ounces of gold per ton. In such a case the settlement would be based on the average of 4.77 ounces per ton in the first instance and 5.25 ounces per ton in the second instance. With gold at $20 per ounce, there would be a difference of $9.50 per ton. Such differences are neither necessary nor desirable.

To appreciate these conditions more fully it must be realized that even with a lot of ore amounting to several carloads the division into these multisamples is usually made after the quantity retained for this division has been reduced to 500 pounds or less. It seems proper, therefore, to criticize severely a sampling method that will not check the multisampling of 200 to 500 pounds of ore closer than 10 per cent. It is also proper to speculate on how much greater a difference might have developed had there been used for the check sample the entire lot of ore instead of the few hundred pounds. Furthermore, it should be borne in mind that even if these original and duplicate samples check absolutely, such checking does not prove the correctness of any previous part of the sampling process.

PROPORTION OF ORE RETAINED FOR SAMPLE.

It is unfortunate that, owing to limited storage room and to the length of time involved, it is not feasible for other than custom sampling plants to retain intact the whole tonnage of every lot of ore until final settlement. With smelting and milling plants, therefore, it is frequently the rule that, during the sampling, the bulk of the ore, the portion rejected, is taken directly to storage bins or beds where it immediately loses its identity through being mixed with other lots of ore, only a small part of the whole being retained for resampling and the settlement of possible disputes. In the case of high-grade ores, or of special contracts, some plants retain the whole lot. However, the prevailing custom is to retain a few hundred pounds in the case of ordinary lots and one-fifth of the lot in special cases. The State of Montana has tried to protect both parties to the sale of ore by enacting a law compelling every sampling plant to retain intact 2½ per cent of every lot of ore until final settlement has been made.

The existence of a considerable difference between multisamples is not everywhere considered a disadvantage. It makes the matter of final settlement susceptible to delicate manipulation by both parties; at times it enables the seller to raise the price somewhat and at other times enables the buyer to pay a few cents less than he otherwise would. As the manager of one of the larger plants has aptly said, "Do not interfere much with these original and duplicate or resampling rules, as that is where we make our money."

The records of some sampling mills show that a system is possible. whereby the ordinary difference between two samples will be well within the limits of the accuracy of the assay, say approximately 1 per cent, and that a greater difference than this will be shown in less than 2 per cent of the cases. To make such results general would mean the scrapping and rebuilding of many plants, but the annoyance and cost of such a change would be returned in many ways. Among the advantages would be that the doubt of the mine superintendent regarding the correctness of his returns would be eliminated, as would the difficulty of the smelter or mill superintendent in deciding whether his losses or gains in recovery were due to his metallurgical practice or to a careless or overzealous sampling foreman. There are few sampling-mill foremen who have not worked overtime trying to develop some system that would enable them to avoid during the current month the blame attached to the sampling department for the loss of metals in the metallurgical operations during the previous month, and frequently such foremen have been unable to locate the cause of discrepancy and have merely passed along the reprimand of the superintendent to the shovelers or other workmen with the remark: "Be careful not to get too much of the 'fines' in the sample; the superintendent says that we ran short last month."

Generally a foreman does not enjoy this state of affairs, and no one who has had experience with this hard-working and loyal class of workmen can seriously doubt that the adoption of more nearly uniform and more accurate methods of sampling will be enthusiastically welcomed by them.

MINE SAMPLING.

Although a discussion of mine-sampling methods is not within the scope of this paper, it is necessary to refer to mine sampling on account of the friction that such sampling may cause at times between buyer and seller. Every buyer of ore knows the difficulty experienced by some mines in obtaining even a fairly accurate sample of the ore before it leaves the mine. There have been many unnecessary resamples, and many uncalled-for diversions of shipments through the failure of the large mill sample to check the small and crude grab sample taken at the mine.

DISCUSSION OF METHODS EMPLOYED.

SAMPLING.

The investigation reported here has been limited to the sampling of the ores of gold, silver, lead, copper, and zinc. The sampling plants visited were in the States of Colorado, Utah, Montana, Washington, Nevada, and California. In all 48 mills, belonging to 23 companies, were examined. At each plant a study of the equipment and practice was made and a flow sheet was prepared. In every instance the heartiest cooperation was shown and every assistance rendered for a thorough understanding of the plant and practice. A lively and most gratifying interest in the purpose of the investigation was shown and a large proportion of the men interviewed expressed a desire that some way might be found of eliminating inaccuracies and uncertainties. In almost every plant some idea that was the result of local experience, or experiment, had been applied that would be of value to ore-sampling practice as a whole, should the information be generally disseminated. On the other hand, in many of the mills were found one or more practices that elsewhere had been laboriously proven to be incorrect. Such information should also become a matter of general knowledge.

The very nature of the business of ore sampling is such that it can have no trade secrets. Costs can be and are guarded as a matter of business policy, but no concern can long exist in the ore-sampling business if the methods it employs are at all mysterious. Therefore much is to be gained by a full comparison of methods and the results of experimental work; it is a matter for each individual in the business to decide to what extent he wishes to cooperate. In accordance with the bureau's policy, it was agreed not to publish names or information that would serve to identify plants visited during the investigation, and in the discussion that follows the intent is to keep within the spirit of this understanding.

The flow sheets which are printed on pages 68 to 92 are intended to show the path of the ore from the car or receiving bin to the final sample sent to the assayer. Each flow sheet does not necessarily represent a separate sampling concern. One plant, for instance, has three distinct methods of obtaining what may be termed the first sample, but adopts a single method, that of coning and quartering, for the final reduction; hence the work of this plant is represented by four separate flow sheets. The method employed seemed to be the only feasible one for properly classifying the various systems.

In the practice of ore sampling there are many variations in the application of general theories, and in the construction and manipulation of the machines. In most of the plants visited these variations are clearly in the line of progress and tend to produce better and more dependable results, but in some plants these variations are distinctly retrograde and lead to inaccuracies. A disinterested observer is more or less in doubt as to whether the plant he visits is attempting, if the plant treats its own ores, to show an apparently better extraction of metals, or, if it buys ores, to assure a safer margin for the ore-purchasing department.

In a number of plants the laws controlling the separation of the finer from the coarser particles seem to be well understood and their application well worked out, but in other plants these vital points are apparently not clearly understood. Long-used methods are continued, because it seems wise not to disturb conditions heretofore satisfactory from the operator's point of view, because of an unwillingness to acknowledge that a new method can be as good or as reliable as an older one, or because of the lack of dependable information on the general principles of sampling. In confirmation of the reasons given for the existence of these variations, it was noted that, although a number of them worked to the buyer's advantage, there were several instances where they were decidedly to his disadvantage.

In presenting the flow sheets and in describing methods the author has attempted to show some of the good points as well as some of the inherent or special weaknesses. Some deductions as to the effects of certain practices on the accuracy of the various sampling methods employed are made and ways of obviating or mitigating these effects are suggested.

Even after an approved sampling system using accurately built machines has been installed and proper rules and regulations have been established, there still remains one disturbing factor—the personal equation of the workmen. Advantage is taken of certain conditions of sampling that may result in hardship to one or the other of the interested parties, whether through the instinct of the workman to protect the interest of the buyer, his employer; or through the ore watcher, who represents the seller, not being inclined to call attention to any operation which would cause an excessive valuation; or through enjoyment of a shrewd bargain causing either party to take undue advantage where possible.

To impress the workman with the fact that perfect fairness to both buyer and seller is the important feature of the work is one of the most difficult problems. It does not belong to mechanics, nor is it entirely moral, but possibly would come under the heading of "Psychology of ore sampling."

It is not the intention or desire of the author to place blame for this condition on any particular buyer or seller of ore, company, superintendent, or group of workmen; neither does he claim that such practices are confined to any particular locality or that erroneous results are intentionally obtained at any particular plant. Decision regarding these matters must necessarily be left to the judgment of the individual in his relation with the plant. However, it is thought proper to discuss the effects of certain practices, whether intentional or accidental, on the accuracy of sampling, in the hope that a clearer understanding of them may work to the mutual advantage of both the buyer and the seller of ores.

THE WEIGHING OF LOADED ORE.

Ore is generally weighed on railway platform scales, which vary in capacity from 60 to 100 tons. The scales are either placed under cover or in the open, and may be near the unloading plant or as much as one-half mile from it. In the past 10 years the practice of weighing has improved greatly. In many districts the scales are tested at frequent though irregular intervals by the Western Weighing Association. These tests are made by the use of cars built especially for this service, and have resulted in general satisfaction. Where these cars are not available, it is necessary to use the 50-pound test weights, of which every plant has 20 to 40. This system is far from satisfactory, for an undetected difference of 20 pounds in a 2,000-pound weight would result in a 2,000-pound error in weighing 200,000 pounds.

It is the general practice to cover the space between the coping and the scale platform with a piece of old belting to prevent accidental jamming of the platform with pieces of rock. However, this practice is by no means universal, for some operators assert that the belting does not exclude the rock and does prevent the constant inspection necessary to insure the free movement of the platform.

In most of the plants visited the weighing is carefully done, but a very bad practice noted in some instances is undue haste. Hasty weighing means that the weights and the rider must be rapidly shifted. In weighing the loaded car, the rider may be thrown far to the overweight position and then rapidly shifted back to a position of balance, the reading being taken as the indicator passes the center on the way up. This will cause the scales, especially if they are sluggish, to register too low a weight. If the tare weight be taken by first throwing the rider to an underweight position, the total difference may be seriously below the correct weight. If the practice is reversed, the net weight will be correspondingly too great.

A practice that has been ordered discontinued by the Western Weighing Association is that of weighing a slowly moving train of cars while coupled together. As the scales must be balanced and read quickly, frequently with the scale beam oscillating rapidly, the possibilities of error due to this practice are apparent. With scales that are too short for some of the longer cars it is customary to weigh separately the load on each set of trucks.

Everything connected with the scales should be kept as free from dust as possible. At many plants the scale beam is situated in a closed room, far from the disturbing effects of the mill vibration, but unfortunately this practice is not universal. An old plan, which possibly is not now in use, was to keep on the rider several ounces of fine ore that was added to or taken from in balancing the scales. An error of several thousand pounds can be caused in the indicated weight of a carload of ore by this system. The weight of the rider is absolutely fixed at the factory and usually a seal is placed where the adjusting material is added or removed. This rider should be kept free from dust or dirt.

When ore is delivered to the plant by wagon, the roadway for a short distance at each end of the scale platform should be level. When the loaded or unloaded wagon is weighed the wagon brakes should be loose and the harness tugs slack. Several hundred pounds may be added to the net weighing by the horses straining in their collars when the loaded wagon is being weighed, or backing into the breeching when the empty wagon is being weighed. The driver and his assistant should occupy the same position during the weighing of the empty wagon as in the weighing of the loaded wagon.

METHODS OF SAMPLING.

Theoretically, the sampling of a lot of ore is a continuous process from beginning to end. In this paper it seems advisable to discuss sampling under two headings, the "outside work," or all the operations up to the point where the sample is to be dried, and the "final work," which includes the drying, grinding, and all bucking-room operations.

OUTSIDE WORK.

The sample resulting from the outside work usually weighs from 25 to 250 pounds and may be obtained by any one of a number of methods. These methods may be divided into so-called hand methods, involving the use of nothing more elaborate than a shovel and wheelbarrow, and mechanical, sometimes called automatic, methods, involving the use of more or less complicated machinery.

HAND METHODS OF SAMPLING.

GRAB SAMPLING.

The simplest form of hand sampling is "grab sampling," used for the most part for sampling large tonnages of low-grade and uniform ores that will not stand the cost of a more exact method. From the pile or carload of ore to be sampled small quantities are taken by hand, or with a small scoop or a shovel, and are thrown into a container or into a pile and held for further treatment. The grabs are taken at random from the pile, or the surface of the pile is laid off in squares by more or less accurately measured lines, and a grab taken at each intersection; or holes are dug at regular intervals in the pile and all or part of the ore so removed is used for the sample; or a grab sample may be taken from the face of a pile of ore as it is being moved from one position to another.

This is the most rapid method of sampling a lot of ore, and, with ore of uniform grade, careful workmen obtain fairly accurate results. It is most valuable when a quick but rough check is desired on the regular sample, as, for instance, to check a possible gross error or to detect quickly whether the regular sample has been salted. The difficulty in getting the proper proportion from a small pocket of fine ore, and the equal difficulty of knowing how much to take from the large lumps, places the whole process at the mercy of the operator's judgment, and his knowledge of the distribution of the values may unduly influence him in selecting the material for the sample.

PIPE SAMPLING.

Pipe sampling can be used for fine material only, and is therefore usually confined to mill products. The simplest form of pipe sampler is a section of 1-inch pipe which is driven or pushed into the material to be sampled, then withdrawn and the contents removed by pounding the pipe. This operation is repeated a number of times until a sample of the desired size results. Another more elaborate form consists of two pipes, one fitting within the other and both having a longitudinal slot and being so arranged that after the pipes have been driven into the ore one can be turned to close the slots before being removed from the ore.

The grab and pipe methods of sampling were not found to be in general use and are so rarely suitable for ores that a more extended discussion of them seems unnecessary.

CONING AND QUARTERING.

Coning and quartering is the oldest and best known method of sampling, and its introduction probably marks the beginning of scientific sampling. An account of its history and development would make an interesting chapter. Its simplest application consists of shoveling the ore into a conical pile, flattening the cone thus formed into a circular cake of larger diameter than the base of the cone by dragging the ore radially or spirally with a shovel, and dividing the cake into four equal sections, called "quarters," by two lines at right angles to each other and passing through the center of the cake. Either two of the opposite quarters, called the reject, are removed by shoveling and taken to the bedding floor or to storage bins. The two remaining quarters are shoveled into a cone, which is flattened and subdivided as before. This process is repeated until a sample of approximately the desired weight remains. The portion retained throughout the operation at the various stages is called the "sample."

Coning and quartering is in use exclusively in 2 of the plants and in combination with other sampling methods in 26 of the plants visited, the general practice being shown in flow sheets Nos. 1 to 7, inclusive.

The advantages claimed for the coning and quartering method are as follows: An expensive plant or complicated machinery is not required; the method can be used where mechanical apparatus is not procurable; it is applicable to all kinds and conditions of ore; it may be used for a small lot of high-grade ore without the loss that might result from putting the ore through a large mechanical sampler; and it often pleases the seller because his ore is in plain sight during the whole operation.

The disadvantages, however, are so numerous that it is difficult to understand why it so persistently survives. It is expensive, for the ore has to be moved many times by shovel and wheelbarrow; much time is lost in the necessarily frequent sweeping of the floor; it offers ideal conditions for salting by the seller; and the slightest carelessness may cause a serious error.

Whenever a lot of ore containing particles of varying size is piled in the form of a cone with a shovel or by falling from a vertical or an inclined chute, a rough separation of the fine and coarse particles is begun and is continued to the end of the operation. This is true even when the ore is finely crushed, although the separation becomes more noticeable as the differences of the sizes increase, as is shown in Plates I, A, and II, A. There is no possibility of uniformly mixing the ore by this method, no matter how conscientiously the method may be carried out. The most that can be claimed for it is that the different sizes are more or less evenly distributed around the vertical axis of the cone. (See Pl. I, B.) The separation of the particles begins with the first shovelful, and after the cone is a foot or more in height the coarser particles run down the sides, the larger ones continuing to the floor. The finer particles remain at or near the apex, while the intermediate sizes lodge on the sloping surfaces and are forced nearer to the floor with each added shovelful. If the ore be coned from a chute or spout, this condition is exaggerated, because the ore flows in a continuous small stream instead of in intermittent and heavier shovelfuls.

MAKING THE CONE.

It may be well to describe more fully the methods of making the cone in order that the reader may clearly understand how this segregation of coarse and fine particles may be exaggerated. First, a point on an iron plate or wooden floor is selected as the center of the cone. The ore is then brought from the bins or cars in wheelbarrows and dumped in one rough pile, or in a number of smaller piles around the central point, or the ore is dumped in a rough pile over the proposed center of the cone and then shoveled back into a ring or into four small piles or cones. The regular cone is then formed by shoveling from the ring or piles (fig. 1). If the shovel is held close to the floor in shoveling any pile of ore, the face of the ore soon reaches its angle of rest. The coarser ore then rolls down and stops a short distance from the base of the cone, while the finer ore slides gently down and remains at the base. The shovelers usually stand between the forming cone and the ring or piles of ore from which they are shoveling. The shovelers may push their shovels into the pile of ore radially or at an angle. If the shovel is pushed into the ore radially, the coarse ore will be found on the heel of the shovel

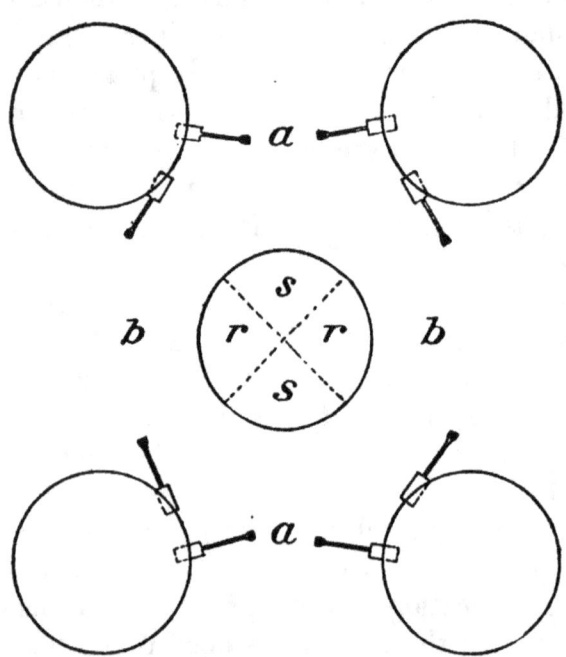

FIGURE 1.—Method of forming cone by shoveling from four piles to a central pile.

and the finer ore on the point; if at an angle, the coarser ore will be found on that corner of the heel farthest from the pile and the finer ore on that corner of the point nearest to the pile. This is clearly shown in figure 1, which represents four piles of ore placed about the proposed central cone. The reject and sample quarters are represented in the figure by r, r, and s, s, respectively. If the shovelers at a, a push their shovels straight into the piles, then when the load is thrown on the apex of the cone the coarse ore will fall from the heel of the shovel into the sample quarter nearest to them, and the fine ore from the point of the shovel will tend to fall in the opposite sample quarter. If the shovelers at b, b load their shovels by pushing them into the piles at an angle, but throw the load squarely on the apex of the cone in the same manner as those at a, a, then the coarser ore will fall into the sample quarters, *s, s,* and the finer into the reject quarters, *r, r.* It might appear that, as *a* and *b* are opposite *a* and *b,* respectively, the selective work of one pair of shovelers would compensate for that of the other; but this is not true, because from *a, a,* both of the sample quarters *s, s* receive large quantities of the coarsest and finest particles of the ore, while *r, r* neverre ceive their full proportion of the coarsest particles. From *b, b,* the bulk of the coarse particles will reach the sample quarters *s, s.* Consequently, while all the quarters receive some of the fine particles, the greater proportion of the coarse particles is thrown into the sample quarters. If the coning and quartering have to be repeated several times and the work is similarly conducted, the effect will be cumulative and be responsible for inaccuracies otherwise unaccountable.

After the cone has reached a foot or more in height an onlooker will often notice that the apex has shifted to one side (see Pl. I, *B*) and that it is frequently over one of the quarters to be rejected. Also there is small probability of the apex being always in the same vertical line, as it usually shifts over the face of the cone, depending on the height of the shoveler or on the force he used in discharging the load from his shovel.

Conditions similar to those described were noted at several plants, although it was not apparent that advantage was being intentionally taken of them. In coning and quartering, workmen usually acquire certain habits, such as marking the quarters in a certain direction, always taking the rejects from certain points of the compass, or following each other in regular routine around the pile or piles, so that a resample may show a surprisingly close check with the original sample and yet both samplings may be inaccurate.

After the bulk of the ore is on the cone the floor is swept and the sweepings added to it. A common mistake made is that instead of carefully adding the sweepings in small portions to the apex of the cone they are sometimes swept up against its base and may be either included in the sample or taken out with the reject.

FLATTENING THE CONE INTO A CAKE.

The result of this operation is a conical pile of ore (see Pls. I *A, B,* and II, *A*) in which the coarser particles are near the base and the finest are near the apex. Vertical sections of such a pile will show plainly the stratification and also illustrate the lack of proper mixing. The importance of the next process, flattening the cone into a cake, is apparent. This is done with a shovel held in an upright position with its face away from the center. The more common method is to start about one-third of the distance between the outside and the center of the cone, push the shovel in to a depth of a few inches, and then drag the ore beyond the base of the cone, as shown in Plates

I, *C*, and II, *B*. As the ore is dragged along, the finer particles sift out and are left behind, while the bulk of the coarse particles is carried to the end of the drag. As this process is continued, each succeeding drag is started from a point a little nearer to the center of the cone and therefore contains a larger proportion of fine particles, so that the tendency is to make a cake having the bulk of the coarse particles on the bottom and the finest on top. This tendency is shown by the view of a cake from which the reject quarters have been removed, shown in Plate II, *C*.

In some plants, as a preliminary to dragging as described above, or sometimes as a substitute for it, the shovel blade is held vertically with its plane nearly parallel to the radius of the cone, and is then churned up and down as the workman walks around the cone. When the workman has made one circuit of the cone, he moves his shovel a little nearer the center and makes a second circuit, repeating the process until he has reached the center. The churning action of the shovel intensifies the segregation of the particles and, after the first circuit, the remainder of the cone is seen to be surrounded by a ring of the coarsest particles. The circular dragging motion is intended to overcome in part the possibility of improper coning, as portions of the ore may be dragged some distance around before being released by the shovel. The principal disadvantage of this modification is the greater segregation and the possibility of leaving more or less of the fine particles in certain quarters.

In both of the systems described it is difficult to uniformly distribute the last few shovelfuls of the finest ore from the apex of the cone. In consequence this fine material may be dragged bodily into some one of the quarters, thereby seriously affecting the result of the sampling.

Another practice is to remove from the center of the cake a portion extending to the floor and a foot or two in diameter. The value of the sample depends on whether this fine material is uniformly distributed all over the cake, or whether it is dragged into the sample or reject quarters. The distribution depends largely upon the personal equation of the workmen. The use of this method increases the proportion of fine material on top of the cake and therefore the difference between the top and the bottom layers and the possibility of error in quartering.

BENCH OR COBB SYSTEM.

A method known as the "bench" or "cobb" system, by which the tendency to segregation is largely overcome, is in use at two plants. It was first applied to certain high-grade ores and has given such satisfaction that it has been continued in use for over a decade.

Instead of first piling all of the ore in a cone, a certain part, sometimes as little as 15 pounds, is coned. This small cone is then dragged out into a layer 1 inch or more in thickness and a similar quantity is coned on top of this layer and dragged out. This is continued until all the ore is in the cake, ready for quartering. This method gives a cake with a surprisingly small amount of segregation and, as there is no large accumulation of fine particles at the center, the effects of accidental error in dragging out the ore are not so serious as in the original method of flattening from a single large cone.

USE OF WOODEN OR STEEL CROSS.

The cross, which was used in the early days of mining in Colorado, was still in use at seven of the plants investigated. The arms, placed at right angles, are of steel or wood, any desired length or width, but of equal size. Crosses of different sizes are used, depending on the quantity of ore to be sampled.

The cross is placed on the floor where the cone is to be formed and the ore is then shoveled onto it, the intent being to throw each shovelful directly over the intersection of the arms. After all the ore is placed on the cone, it is flattened by the usual method into a cake as thick as the cross. The cross may also be used for quartering after the cake has been formed. This is accomplished by placing a steel cross on the cake in such a manner that its intersection comes at the center of the original cone and then driving it down to the floor; the rejected quarters are then shoveled from between the arms. The special advantage of this device is apparent from the following description of the quartering.

The flattened cake, if the cross is not employed, is usually marked with a straightedge or, if the cake is small, with the handle of a shovel. In order that every quarter may contain its proper proportion of fine and coarse particles, it is obviously necessary that these lines should intersect at the center of the original cone. Sufficient care is not always given this important step, and it is not unusual for these lines to be marked 2 or more inches to one side of the center. In this way it is possible to take advantage of any known shifting of the apex of the cone. The reject quarters are now shoveled from the cake into wheelbarrows and taken to the bin. After the bulk of the reject is removed as closely as possible to the marked lines on the cake, the shovel is either driven down vertically from these lines to the floor or drawn from the top to the bottom at an angle approximating the angle of rest, the ore thus removed being added to the reject. Segregation of the particles will be most apparent at this point, and the thicker the cake the greater will be the difference between the sizes of the particles at the top and at the bottom, as shown in Plate II, C. The workmen can not prevent the face of a cut seeking its angle of rest, and the result is that the finer particles at the top usually fall or are dragged into the reject space and are consequently swept up and carried away, a corresponding amount of coarser material being left in their place. Although this amount is not large, it may, with high-grade ore, cause a serious error.

With the open system, the workman can never be sure that he is taking from these division edges more or less than he should. He also has to exercise his judgment as to whether he should take all the fine particles that fall to the floor, or whether he should dip into the layer of coarse material that lies close to the floor. This error may be minimized by the use of the steel cross, as previously described. Whether the cross is used at the time of coning or for the quartering only, it is an improvement over the open system, and its general adoption is a matter worthy of consideration.

Throughout the whole operation, from the time that the ore is first dumped on the floor to the time that the reject quarters are removed, the tendency is to reject an excessive proportion of fine particles, thus leaving for each succeeding sampling a constantly increasing proportion of coarse material. This tendency can in a measure be overcome by alternately taking for the sample the ore in the usual reject quarters. Inasmuch as the first cone is the largest, its proportionate error, due to segregation, is probably greater than that of the succeeding cones. Therefore, whether the reject or the sample contains the greater proportion of fine particles depends on which one is removed in the first quartering. All operators do not recognize the importance of this segregation, and occasionally a plant permits the seller of a lot of ore to specify, during the sampling, which quarters are to be taken for the sample. Where this is done, an intelligent and attentive watcher may be of considerable value to his employer.

The custom noted in a few plants of removing additional slices from the sample quarters left on the floor, when these quarters are found to contain more ore than is desired for the final sample, and yet is too small to be again quartered, is of doubtful accuracy. If this slice is taken radially, there is less objection to it, but it is usually taken on a line parallel to the already marked radial edge, so that the proportion of fine material to coarse material is increased beyond the correct ratio.

This will be made clear by reference to A and B in figure 2, which represent by full lines the sample quarters of the same cake after they have been reduced by both of the methods outlined in the preceding paragraph. As the finest ore is at the center of the cake, and the ore becomes progressively coarser toward the edge, it will be apparent that the ore lying within the inner circle averages finer than that lying between the inner circle and the circumference of the cake, and that the proportion of fine to coarse ore will be the same in all radial slices throughout the cake. In B the slice *defg*, taken parallel to the radial edge *fg*, contains the same amount of ore as the radial slice *abc* in A, but has a greater proportion of its area lying within the inner circle, and therefore *defg* would contain a larger proportion of fine ore than *abc*. The black parts in the two diagrams have been calculated mathematically to contain the ·same amount of ore, and the shaded part in B represents the discrepancy between the correct proportion of fine ore to be rejected and the proportion that would be rejected if the slice were taken as shown in B. This discrepancy varies according to the size of the slice taken.

FRACTIONAL SHOVELING.

None of the plants examined used the fractional shoveling method of sampling exclusively, but all the plants using the coning and quartering method, and six plants in which mechanical methods were used in combination with hand methods, use fractional shoveling at some point in their operations. It is more commonly used during the unloading of the ore in order to save the expense of passing an entire lot through the mill or, if the ore is to be smelted, in order to put it on the beds in as coarse a condition as possible. Unquestionably it is a very convenient method, especially in unloading a lot of ore that has been previously crushed, or has been sampled at a custom sampling plant before its arrival at the reduction works.

Fractional shoveling has many advocates who claim for it greater accuracy than is possible with the coning and quartering system and

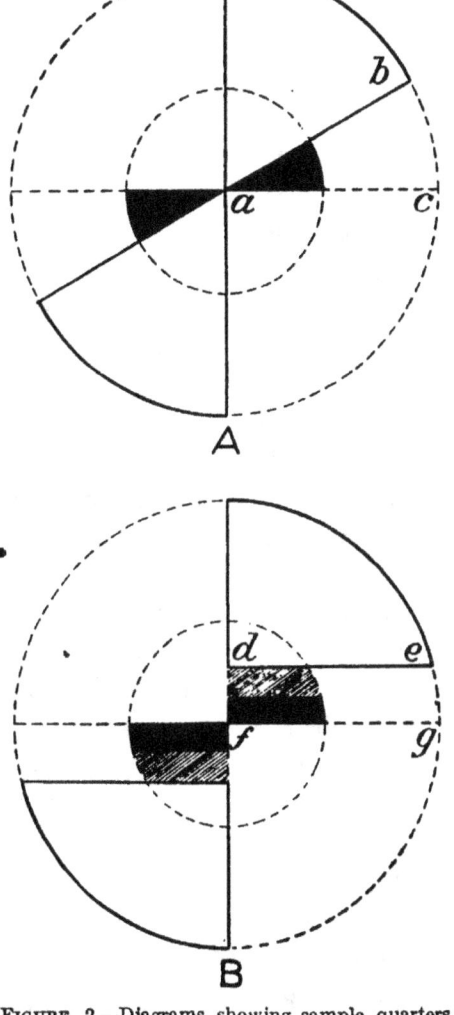

FIGURE 2.—Diagrams showing sample quarters reduced by slicing. A, Radial slicing; B, parallel slicing.

offer figures to show close checking on samples and resamples. If, however, the ore has been coned or placed in a regular pile before shoveling, the same evident sizing exists and opportunities for careless or objectionable shoveling seem as great and, in certain instances noted hereafter, greater than in the coning method. Therefore, except for the convenience and possibly greater economy in handling, it offers little or no improvement over coning and quartering.

When ore received by rail is sampled by fractional shoveling as unloaded, the car is placed over or near the storage bin, or beside a conveyor belt, or near a platform; in the last event the ore is taken to the storage bins in wheelbarrows.

The portion reserved for the sample varies from every second to every tenth shovelful, which is thrown into a special wheelbarrow and delivered to the sampling floor, or may be thrown into a separate pile in the original car. The reject is delivered to the storage bin. Occasionally the first sample, while still on the car, is again shovel-sampled to reduce its bulk and the car switched to a point near the sampling floor and the sample unloaded.

DISADVANTAGE OF FRACTIONAL SHOVELING.

One serious objection to sampling by this method during the unloading of the ore is that the work is generally done by the cheapest labor and frequently by contract. In either case speed is the principal object. Moreover a contractor will not criticize his men for careless work or inaccurate counting of shovelfuls. The usual system is to instruct each man to count his own shovelfuls, but, in one instance of contract work, the man in charge counted in a loud voice the shovelfuls thrown out by a certain workman. When the proper number was reached he called "Sample," when every workman in the car was supposed to throw his next shovelful into the sample pile, regardless of the number he had previously thrown into the reject. No uniformity was noted in this work, as is illustrated by the following records, which were made at three different plants.

In each instance it was the intention to have the tenth shovelful taken for the sample. At the first plant the sample shovelful was consecutively the second, thirteenth, fourth, sixth, twelfth, tenth, ninth, sixth, tenth, third, and eleventh, averaging 1 shovelful in 7.8 shovelfuls. At the second plant the record was the tenth, eleventh, and twelfth, averaging 1 in 11 shovelfuls. At the third plant, there were four men in the car, two working together at each end; each shoveler was also attending to the sample, placing the reject into one wheelbarrow and the sample into another. The following table shows the number of shovelfuls put into the four reject wheelbarrows by each of the two shovelers and the number of shovelfuls put into the sample wheelbarrow by each workman while loading the reject barrow:

Record of shovelfuls put in reject and sample.

	Shovelfuls to reject.	Shovelfuls to sample.
First wheelbarrow	15	0
Do	12	2
Second wheelbarrow	13	1
Do	13	1
Third wheelbarrow	16	1
Do	14	2
Fourth wheelbarrow	14	0
Do	15	2
Total	112	9

Ratio of total sample shovelfuls to total shovelfuls, 1 in 13.3.

CONDITIONS IN SHOVELING.

Fractional shoveling is used at various plants as a part of the regular method of sampling after the ore has been subjected to one or more crushing operations. When this process is employed, the ore is thrown into a cone or regular pile, either on an iron plate or a smooth floor. The workmen then take shovelfuls from the base of the pile close to the floor, throwing the sample shovelfuls on a separate pile for further treatment and the reject shovelfuls into a wheelbarrow to be taken to the storage bins. After the first shoveling the floor is swept and the operation repeated as many times as required to reduce the sample to the required size.

The following discussion will aid in a better understanding of some of the conditions existing in shoveling from a pile of ore. If a shovel is pushed straight into a pile of ore, an imperfect pyramid of ore forms on it as it is withdrawn. As the ore on the shovel seeks a new slope, it starts sliding down the inclined surface, and as the coarse ore has the smaller angle of rest it runs ahead of the fine ore associated with it. If the shovel is overloaded, the coarse ore falls to the floor, leaving an excess of fine ore on the surface of the load. If the shovel is partly loaded, some coarse ore falls from the front and sides and some rolls down and collects on the heel of the shovel. Whether the workman takes consecutive shovelfuls at the same point—that is, on the same radius of the cone—or advances a step around the cone with each shovelful, it is possible for him to reject regularly and intentionally too great a proportion of coarse ore and take for the sample too great a proportion of fine ore. As the shoveling proceeds, the coarse ore that falls down the face lodges a short distance from the base of the pile, and when it has accumulated in sufficient quantity it may be scooped up as a shovelful for either the reject or sample.

A new workman, regardless of whether fractional shoveling is done in the car or on the sample floor, generally takes a larger load on his shovel for the sample than for the reject. This is probably due to the fact that the sample shovelful marks the end of a period and is more impressed on his mind. As he is cautioned against this tendency and grows more experienced, he more often errs in the other direction, and takes too small a sample. As the center of the cone is approached, the proportion of fine ore increases, and if smaller shovelfuls are taken for the sample at this point than were taken on the outside of the cone a considerable error will result.

Inaccuracies in shovel sampling are difficult to detect during the work, and only long practice enables the watcher to decide that any individual workman is doing careless or intentionally inaccurate work. With more than one workman at the pile of ore the watcher's attention is necessarily divided, and, as every shovelful is immediately thrown into a general pile and its identity lost, no comparison can be made of any two or more shovelfuls. Moreover, an experienced shoveler may, to the utter confusion of the observer, change his tactics from time to time, especially if he thinks that he is being watched.

EFFECT OF RUNNING ORE FROM A CHUTE.

One system in use in a few plants increases the possibility of errors in fractional shoveling through the manner in which the ore is piled before and during the shoveling. In this system, as the ore is crushed or rolled, it drops directly to a chute inclined at an angle of approximately 45°. The bottom of this chute generally ends about 3 feet above the sampling floor and rests upon a vertical iron plate placed at right angles to the chute. As the ore rolls down the chute, the varying speeds of the different sized particles causes a rough separation, the larger particles rolling along freely and being discharged at a greater distance from the vertical plate. The smaller particles move more slowly and, not having so great an impetus, drop directly under the chute onto the floor or pile. The intermediate sizes fall at points between these extremes. As the pile of ore increases in height and its top approaches the bottom of the chute, it will form a semicone and the coarse particles will continue to run down the side farthest from the chute, while the fine particles will spread out upon each side and near the vertical plate.

As the sampling plate does not, as a rule, hold a large quantity of ore, it is customary to begin fractional shoveling at about the time that the apex of the pile has reached the base of the chute, thereby causing ideal conditions for the continued separation of coarse and fine particles. The appearance of such a pile of ore is sufficient to convince one of the possibility of error when a sample is removed from it by fractional shoveling.

Figure 3 illustrates the condition described. In the figure a is the inclined chute, $b\,e$ the vertical plate, $b\,e\,d\,f\,c$ the pile of ore, f, g, and h the positions of the shovelers, and i and j the sample and reject wheelbarrows. The line $c\,d$ is the assumed boundary between the coarser and finer particles. Usually a workman in shoveling will assume the most convenient position. If he stands at f he will shovel first into wheelbarrow i and then into wheelbarrow j from about the same part of the pile; but as few men can shovel both right and left hand, the usual position is either at h or g. If the workman stands at h, the wheelbarrow j will receive its load from the neighborhood of d, but the workman will take a step forward and load wheelbarrow i from a point near f. If he stands at g, he will load i from near c and j from near f. In either case the wheelbarrow more distant from the workman invariably receives too great a proportion of the coarser particles, and unless the difference between the value of the coarse and the fine particles is negligible, the ultimate effect depends on which wheelbarrow is taken for the sample.

In the past the seller has taken and probably is still taking an occasional advantage of hand sampling plants and even mechanical sam-

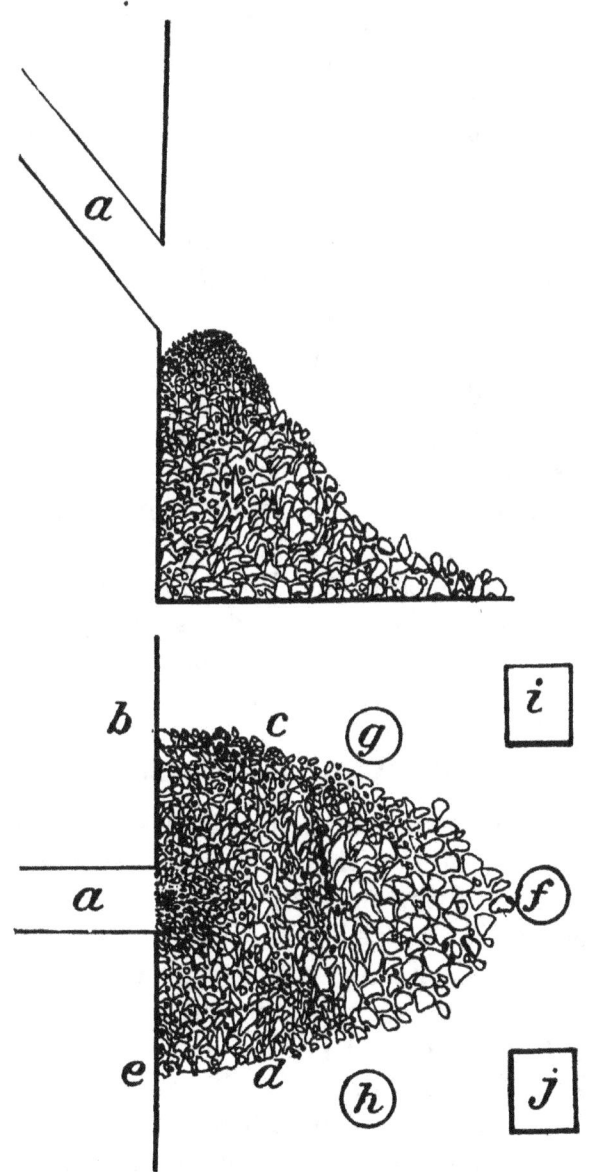

FIGURE 3.—Diagram showing segregation caused by fractional-shovel sampling from cone made under an inclined chute.

pling plants that discriminate against the fines in their sampling. The buyer should have a full knowledge of his system of sampling and of how consistently certain practices are followed in its operation. A few examples will illustrate some of the advantages that the seller may take.

SAMPLING CARS LOADED WITH WHEELBARROWS.

In loading railroad cars at the mine or custom sampling plant by wheelbarrows each load is a rough cone, and it is often so dumped into the car that the apex of all the loads thrown against the side will fall close to the intersection of the side and the floor of the car. If in fractional shoveling of such a carload of ore it happens that the custom of the buying plant is to have the men begin shoveling near the sides of the car, so as to be shoveling near the center when it comes time to take the sample shovelful, the sample will not contain a proper proportion of the fines. Again, if the ore be coned with the idea of shifting the apex to the reject quarter, or if the fines be partly eliminated during the quartering, or if a condition such as shown in figure 3 exists, a similar lack of fines will exist in the sample. If the seller is convinced that no amount of complaint or watching will overcome these conditions, he may, either in his mine or sampling plant or in a railroad car, accumulate a quantity of finely crushed low-grade ore and also a quantity of high-grade coarse ore which he mixes together and ships as one lot, instructing his watcher to absent himself during the sampling of this lot and let the buyer do the rest. The buyer, who has bedded the reject from such a lot, has no protection through resampling, as the grosser errors occur during the earlier stages of the operation. His only warning may be the suspicious satisfaction of the seller, and his only protection may be an occasional test screening of the shipment. Obviously, his best protection would be to install a system that discriminates neither one way nor the other.

MECHANICAL SAMPLING.

In order to avoid many conditions that seem unsatisfactory in all methods of hand sampling, there are in use various devices that are placed in the line of the moving stream of ore and automatically divert for the sample a fixed proportion of the stream. These are known as mechanical or automatic samplers.

Many advantages are claimed for the mechanical system of sampling. Except for the cleaning of the machinery, the only hand labor required is for unloading the ore, and even this is avoided if the ore is delivered in dump cars. The operation is continuous, and if the plant is properly managed the sample may be placed on the drier within a few minutes of the time required by hand methods for merely delivering the ore to the sampling floor for the first coning. The ore is not rehandled nor is it stored for various lengths of time near other samples from which it might easily be salted. The unmixed condition of the ore is less of a factor, and if the machine and general plant are correctly designed and constructed the work can be done at a much lower operating cost. With simple and proper precautions the susceptibility to error and manipulation existing in hand methods is eliminated.

An interesting fact is the universal use of the mechanical system by custom sampling plants. The nature of their work is such that these plants are in a position to compare their sampling systems with those of every other plant to which they forward ore. These comparisons are not only yearly or monthly averages but are also those of individual lots or mixtures of several lots forwarded as one. A well-conducted custom plant makes a study of comparative returns from different smelters and mills and, to this end, frequently makes shipments of what are termed "split lots" of the same ore to two or more smelters or mills. It is not an uncommon practice for custom plants to resample several consecutive portions of the same lot of ore, which are then sent to as many different smelters or mills merely for a comparison of results. They are also constantly studying any change made at any plant with which they have business relations. For these reasons it is probable that a custom plant shipping regularly is better posted as to the reliability of certain methods than is the purchasing plant using them.

The custom plant does not buy all the ore that it samples, but usually does a large proportion of its business in what is termed "sampled-in-transit" ore; that is, ore which the mine sends to the custom plant to be sampled as a check on the sampling that is to be done later at the purchasing mill or smelter. Therefore, the preferable system for the custom plant would seem to be that which is most reliable under all circumstances.

Among the objections to mechanical methods are the greater initial cost of the plant, the heavy cost of renewals of machinery, the difficulty of cleaning thoroughly, the danger of flying dust from a dry, high-grade ore, and, more serious than all else, the fact that with an unscrupulous operator, machines may be and have been so built or arranged that a correct sample is impossible. Too much stress can not be laid upon this feature of the mechanical methods. The combination of an incorrectly constructed machine and improper delivery spouts, as is described in a subsequent page, is a more serious cause of error, than any hand method heretofore discussed. A cone may be improperly formed or the shoveling improperly done but, in both these operations, conditions may be inexplicably changed at any time and, on account of a surprising average of errors, a correct sample may result; whereas, with an improperly designed or adjusted machine and a discriminating delivery, there can be no hope for anything but error. This possibility and its application by many sampling plants in the past has given rise to most of the opposition to the general use of mechanical systems but, fortunately, exposure and agitation have made such practices almost as unsafe as the occasional placing of screens in the bottom of the elevators.

Devices for mechanical sampling operate on either of two distinct principles which separate them naturally into two classes—stationary devices which continuously divert certain fixed sections of the stream of ore for the sample, and moving devices which are so operated that during several fixed periods per minute, they divert the whole of the moving stream of ore for the sample. These devices are commonly and more graphically described as those taking part of the stream all the time, and those taking all of the stream part of the time. The first class of the mechanical samplers now in use is represented by the whistle pipe and the bank or combination riffle.

STATIONARY DEVICES.

WHISTLE PIPE.

The whistle pipe, in combination with coning and quartering, is in use in a number of plants which are under one management, as is represented by flow sheet 52 (p. 91).

Figure 4 is a sketch of this device. The front of the rectangular housing, $e f g h$, is removed to show the whistle pipe, $a b c d$, which is a vertical iron pipe with five notched openings cut halfway through the pipe, as shown at i, j, k, l, and q, each being placed at an angle of 90°, measured horizontally, from the one immediately above. In these notches are rectangular pieces of steel, $m n o p$, so placed that the top edge forms a diameter of the pipe. Above every notch is placed a cast-iron liner, two of which are shown at t and u, so shaped that the lower end is smaller than the upper, to collect the ore in a smaller stream just before it strikes the dividing edge. When the ore is delivered to the top of the whistle pipe, it falls on the first diverting partition, which causes approximately half of it to leave the pipe and fall into the reject bin. The other half continues down the pipe until it strikes the second partition, set at right angles to the preceding one, where it is similarly divided. This process continues through five diverting partitions, so that the sample finally leaving the pipe represents one thirty-second of the original lot. The rejects from the five openings are discharged into the housing and are diverted by the chute $r s$ to the reject receptacle.

The advantages of this device are its low original cost, economy of operation, and the simplicity and rapidity with which it reduces the ore to such a small proportion of the original amount. A disadvantage is that there is no possibility of recrushing the ore until it has been reduced to one thirty-second of its original bulk. This makes it necessary to crush the whole lot to whatever degree of fineness is required by this small sample. As it is rather unusual to change a set of rolls or a screen for lots of small tonnage, it is probable that the crushing limit is that of the average large lot. A maximum size of particles which would be safe for a 100,000-pound lot would tend to cause an error in a 20,000-pound lot. Another possible source of error is that the rectangular piece of iron forming the partition is inserted on an angle rather than in a vertical position. As this partition wears, the dividing edge will recede toward the outside of the pipe and cause too large a proportion to go into the sample. If the feed is uniform, the error resulting may not be serious, but if the stream of ore has a tendency to segregate, a recession of the dividing edge may cause a serious error. Inasmuch as the device must be housed to prevent the escape of ore and dust from the reject, there is little or no opportunity for inspection during sampling, so that a temporary clogging at the sampling partitions could readily remain undetected for varying periods of time, thus forcing an undue proportion of the ore into either the sample or reject. The cast-iron liners tend to wear unevenly, the greater wear taking place where the coarse ore strikes, thus establishing lines of flow that might result in an improper division of the sample.

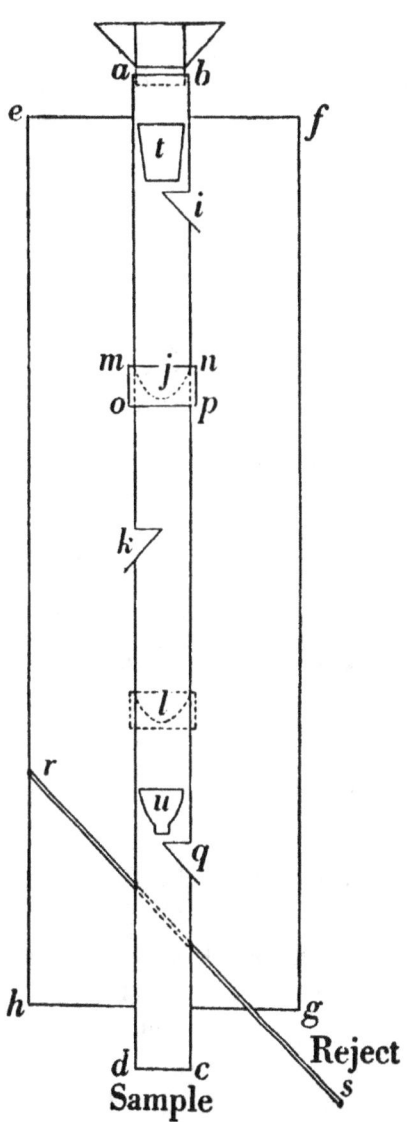

FIGURE 4.—Sketch showing elevation of whistle-pipe sampler.

BANK OR COMBINATION RIFFLE.

The bank or combination riffle is in use in connection with the Vezin sampler (see pp. 35–39) in four plants, as shown in flow sheets 10, 11, 12, and 19. This device consists of five riffles set in one frame, the top riffle being placed over two lower ones, which are in turn followed by two more set below them. Figure 5 illustrates the lines of flow through this device. The ore is fed from a chute or hopper to the first riffle, *a b*, where the stream is divided into a number of smaller streams, of which every other one falls on one side and the rest on the opposite side of the riffle. From the spouts on both sides of the first riffle the streams of ore impinge on inclined iron aprons, k and k, by which they are diverted to the two riffles $c\,d$ and $e\,f$, smaller than $a\,b$. Four sets of streams flow from these two riffles, each set representing one-fourth of the original lot. Two of these unite and fall into the reject receptacle; the other two sets are diverted to two aprons, m and m, and the two riffles $g\,h$ and $i\,j$, where they are again divided into four sets of streams. Two of these sets unite and fall into the reject receptacle; the other two, each being one-eighth of the original lot, may be united or may be kept separate as original and duplicate samples. If the sample so produced is larger than desired, it may, with or without further crushing, be passed through a similar set of riffles, or it may be elevated and passed through the same set. This repetition may be continued until the desired weight is obtained. The riffles may be built in a rigid frame, or they may be hung with rods so that a workman may swing one or more of the sections across a falling stream of ore.

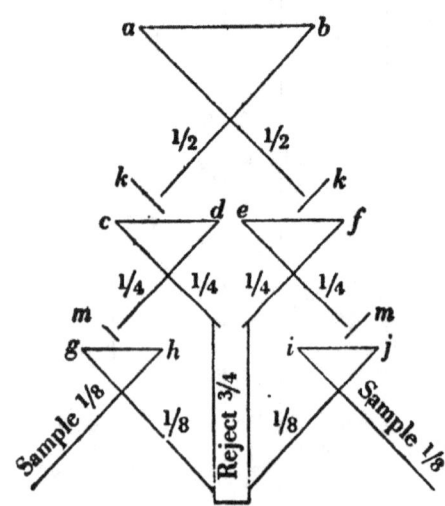

FIGURE 5.—Lines of flow in bank or combination riffle sampler.

Among the advantages claimed for this device is the fact that it divides the original stream of ore into a large number of smaller streams before separation, thus avoiding the error possible in coning and quartering where only four cuts are made before separation. Further, it is easily operated and, as the ore is generally finely crushed when it reaches this point and is fed to the device without a long drop, the wearing of the dividing edges will be slow.

On the other hand, its construction is not simple, and great care is required in constructing the dividing partitions and preserving them in proper condition. There is danger of the partitions becoming clogged by damp ore, pieces of wood, or cloth, and as they are not easily accessible for examination they may remain clogged and part of the riffle may for long periods of time take no sample or else an undue proportion of the ore may pass into the sample. In order to guard against this tendency, the riffles are frequently tapped with a hammer and may be seriously bent out of shape. The great disadvantage of all devices of this type is that, under nearly all conditions, a falling or sliding stream of ore acquires certain tendencies that are developed by the conditions surrounding its previous handling. For instance, if one side of a delivery chute is lower than the other or if there is a decidedly low place anywhere in the bottom of the chute, the fine ore will seek that low point. If there are several such places or depressions, all will form lodging places for the sliding fine ore. If there is a straight delivery to the chute, the coarse ore will be more or less evenly distributed in its flow, but if there is an angle at any point in the delivery, the coarse ore will also be found to have regular lines of flow.

The uneven wear of the face of a set of rolls tends to deliver the crushed ore in separate streams and, at the same time, to segregate the coarse and the fine ore. An angled spout, the effects of which will be more fully referred to later, causes an evident segregation. In fact, any number of conditions may exist that make the delivery of a stream of ore anything but uniform. Therefore, it is possible, and highly probable, that one or more of the divisions may intermittently or regularly receive an incorrect proportion of the fine or coarse ore and the cause may be exceedingly difficult to detect. Correct samples may be nevertheless obtained on account of the compensating averaging of errors.

In the bank riffle the ore that slides down any one of the small spouts tends to separate into different sizes as it impinges upon the small diverting apron. The finer and damper ore adheres to the apron surface for an appreciable length of time and, when it is finally dislodged, it drops into the nearest division in the succeeding riffle. However, another pile immediately begins to form, so that there is built up a series of pyramids which causes the rolling coarse particles to be diverted to both sides and consequently into certain divisions in the following riffle. The first diversion may cause a small and hardly appreciable error, but as the ore passes through three sets of riffles and may be repassed one or more times, there may be a cumulative error which, especially in sampling high-grade ore, may seriously affect the final sample. This tendency is so well recognized that riffles are sometimes given a swinging motion, as has been stated above, in order to break the lines of regular flow. However, as this correction in the regular process generally depends on the memory, convenience, or whim of the workman, it is at best spasmodic.

MOVING DEVICES.

The second type of mechanical sampling devices is represented by the various time samplers that have either a rotating or an oscillating motion. These samplers are so constructed and operated that, during one-twentieth to one-fifth of the time of a single rotation or oscillation, the entire stream of ore is diverted for the sample, and during the balance of the period the entire stream of ore falls into the reject receptacle.

This type has certain special advantages in addition to those already ascribed to mechanical samplers. With proper construction and operation, the manner of delivering the ore is theoretically unimportant. Even though the ore be screened and the fine and coarse delivered to the sample spout from opposite sides of the chute, no error should occur, because the machine cuts across the entire stream, consequently both the sample and reject are under every section of the stream for the same proportion of the time. Owing to the sample spouts being wider than is possible in the riffle system, there is less danger of clogging and the entire machine can be made more readily accessible for cleaning and repairing. It is as easy to build a "crooked" as an honest machine.

The more common examples of this type in use in the plants visited

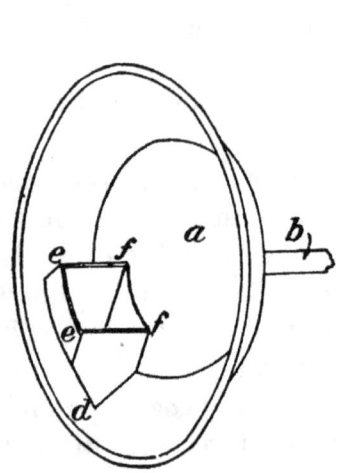

FIGURE 6.—Sketch of Snyder sampler; *a*, disk; *b*, shaft; *e f*, cutting edges of sample spout; *e f d*, side of spout.

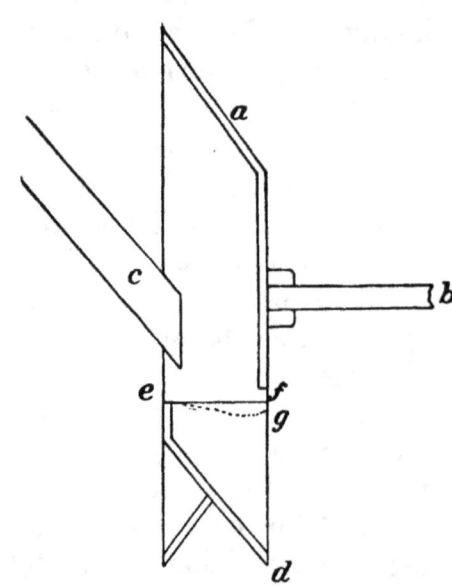

FIGURE 7.—Ideal section of Snyder sampler.

are the Snyder, the Vezin, the Chas. Snyder, the Brunton vibrating, and the Brunton oscillating machines.

SNYDER SAMPLING MACHINE.

The Snyder machine is in use at one plant, that shown in flow sheet 21. This machine (figs. 6 and 7) consists of a cast-iron plate *a*, revolving in a vertical plane on the axis *b*. The spout that passes through the plate receives the sample portion from an inclined delivery chute. When the sample spout is not beneath the ore chute, the ore impinges against the plate and is thereby thrown back into the reject receptacle. Being made of cast iron, these machines have no easily bent or twisted parts, are easily accessible for cleaning and keeping in repair, and, under proper conditions, should give a correct sample. The greatest danger of inaccuracy lies in improper construction. As is shown in figure 7 the cutting edges, *e f*, of the sample spout should be parallel and preferably perpendicular to the plate, and the sides of the sample spout should be in planes passing through the center of rotation. With this construction, the opening in the spout will become wider as the edges wear down, but at the same time it will acquire a longer radius of rotation and will consequently continue to cut the same arc in the same period of time. On the other hand, should these sides vary from the construction described, as, for example, should they be parallel or converging toward the bottom, the wearing of the edges will not cause a corresponding widening of the spout, and the greater the wear the less will be the proportion taken for the sample at that point. If the sides spread more than they should at the bottom, the wear will cause an increasing proportion to be taken in the sample. As the delivery chute is at an angle to the horizontal, the coarser particles will be delivered nearer to f and the finer particles will fall near e, and therefore, the edges of the sample spout will show a greater wear at f, as shown by the dotted line $e\,g$. An error may also be caused by some of the fine ore that ordinarily would fall into the sample spout, adhering for a time to the plate a, and then dropping into the reject, or some of the fine ore that ordinarily would fall into the reject dropping into the sample spout. This occurs principally with damp and sticky ores.

Owing to the slowness of rotation, it may be questioned whether the sample is taken frequently enough. This error is largely overcome by having two, three, or four sample spouts cut in the disk, though this increases the probability of error through sticky ore, as suggested in the preceding paragraph.

VEZIN SAMPLER.

The Vezin sampler is in use in 25 of the plants inspected and is used in combination with almost every present method of sampling.

The details of the construction of the machine have been changed to suit the ideas of many managers of many plants and some of the designs bear only a general resemblance to the original drawings. However, the principle of the machine remains the same and what is said regarding the original form applies to all variations. The Vezin sampler in its simplest form is shown in figure 8.

The shaft, which revolves in the direction of the arrow, carries the sampler with it. The sampler spouts s and s pass beneath the delivery chute c, and deliver the sample part to the center of the machine, whence it passes out through the spout at the bottom. When the sample spouts are not passing beneath the chute, the ore falls into the reject compartment r, which is not shown in the figure.

This device has many distinct advantages. It can be so erected as to be easily accessible for cleaning and examination during sampling; it can be built to all sizes; by the addition of extra spouts any proportion desired may be taken for the sample; two machines may be operated side by side in such a manner that a duplicate sample may be taken at the same time as the original; and when correctly built and the ore properly fed to it, should unquestionably take a correct sample.

Among its disadvantages are the amount of headroom required, the liability of damp ore piling up in the sample spouts, and the tendency of the long upper edges of the sample spout to become bent and uneven. It is not feasible to operate this device at very great speed, and consequently, with the usual one or two armed devices, the samples are taken less frequently than is desirable. This defect can be overcome by increasing the number of sample spouts, but in only one instance was a Vezin sampler noted with more than two spouts. Its great disadvantage, almost enough to condemn it, lies in the ease with which it can be constructed and operated to take almost any kind of a sample desired. Modification of its construction so slight as not to be noticeable when the machine is in motion and only determinable by careful measurements taken with the machine at rest will cause an appreciable error, and if to this is added the error due to an incorrect delivery chute, there would seem to be no limit to the possible error. Moreover, the machine may be so built that, as first erected, the sample spout will show the proper alignment and yet, through the wearing down of the cutting edges, a condition soon arises that is the equivalent of incorrect construction.

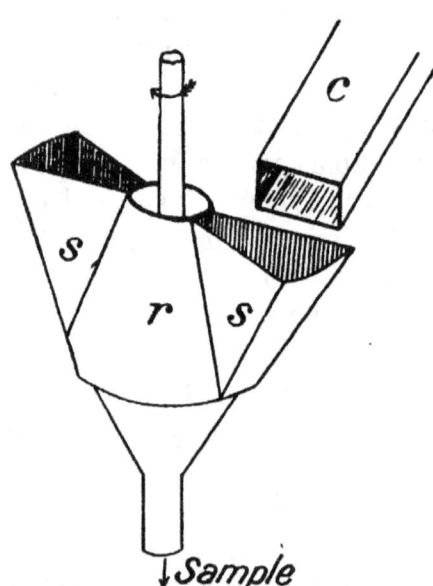

FIGURE 8.—Sketch of Vezin sampler. c, chute; s, s, sample spouts; r, reject.

As this device was never patented, conditions of its manufacture and use could not be controlled, but the inventor, H. A. Vezin, continually urged certain necessary conditions. He insisted that the sides of the sample spout should be vertical, that the top edges should be radii of the circle of rotation, that the feed chute should be inclined at an angle of approximately 58° to the horizontal and should deliver the ore in a sheet parallel to the edges of the spout as they pass beneath it, and that the sampler should be so rotated that the speed of the sample spout should approximately equal the horizontal speed of the ore when it reaches the spout. The latter recommendation was supposed to eliminate the bounding around of the large pieces of ore, to lessen the danger of large chunks jamming between the edges of the chute and the spout, and incidentally to lessen the wearing of the cutting edges. Unfortunately, however, these recommendations are not always adopted. Some of the earlier machines were even constructed with rectangular sample spouts, and, further, the ore was fed over an inclined chute delivering toward the center of the device. As these radical errors in construction were soon understood by the ore seller, variations less apparent were adopted and to-day, to determine whether the sample spout is not nonradial, it is necessary to take careful measurements while the machine is at rest.

In order to more clearly understand this condition, reference should be made to figures 9 and 10, representing horizontal drawings of five Vezin samplers. In the figures *e f* is the outside line of travel of the spout, *g h* is the inside line of travel of the spout, the center of the carrying shaft is shown by a cross, *a b c d* is the plan of the top of the sample spout, and *i j* is the bottom edge of the inclined chute delivering ore to the sampler.

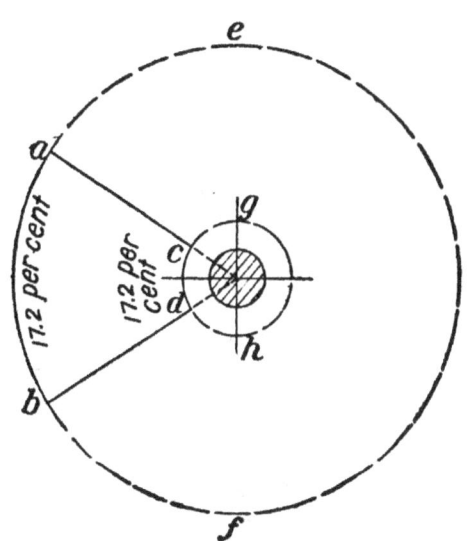

FIGURE 9.—Outline of Vezin sampler spout correctly constructed.

Figure 9 shows the spout correctly constructed, with both *a b* and *c d* cutting the same angle on the respective circles. The lines *a c* and *b d* are radial; that is, a continuation of them passes through the center, so that no matter what form the delivery chute may be the spout should take a correct sample. Figure 10 shows spouts incorrectly constructed, with the inside edge *c d* of the spout wider than it should be and with the continuation of the lines *a c* and *b d* meeting beyond the center of revolution. This causes the percentage taken by the spout to vary from its inside to its outside edge between the figures noted on each drawing. In figure 10, *C*, the line *i j* is that of a delivery chute, and the figures 10.3 and 11.1 are the percentages that would be taken by the sample spout between the points *i* and *j*, and the line *i′ j′* is that of another delivery chute, and the figures 9.4 and 10.3 are the percentages that would be taken by the sample spout from this delivery chute. Figure 10, *D*, represents a sampling machine having the inside edge of the spout narrower than it should be, with the percentage varying from 3.9 on the inside to 11.4 on the outside edge. The spouts would take from the ordinary delivery chute on the line *i j* percentages varying from 9.7 to 8.65.

The differences in percentages noted between the inside and the outside of the delivery chute might be small enough to be neglected if it were not for the fact that the ore does not ordinarily flow from the chute in a uniform stream.   As noted before, the coarse and fine ore will take different lines of flow and the natural tendency to segregation may be greatly increased by an accidental or intentional tipping of the chute.   For instance, if the chute should be so tipped

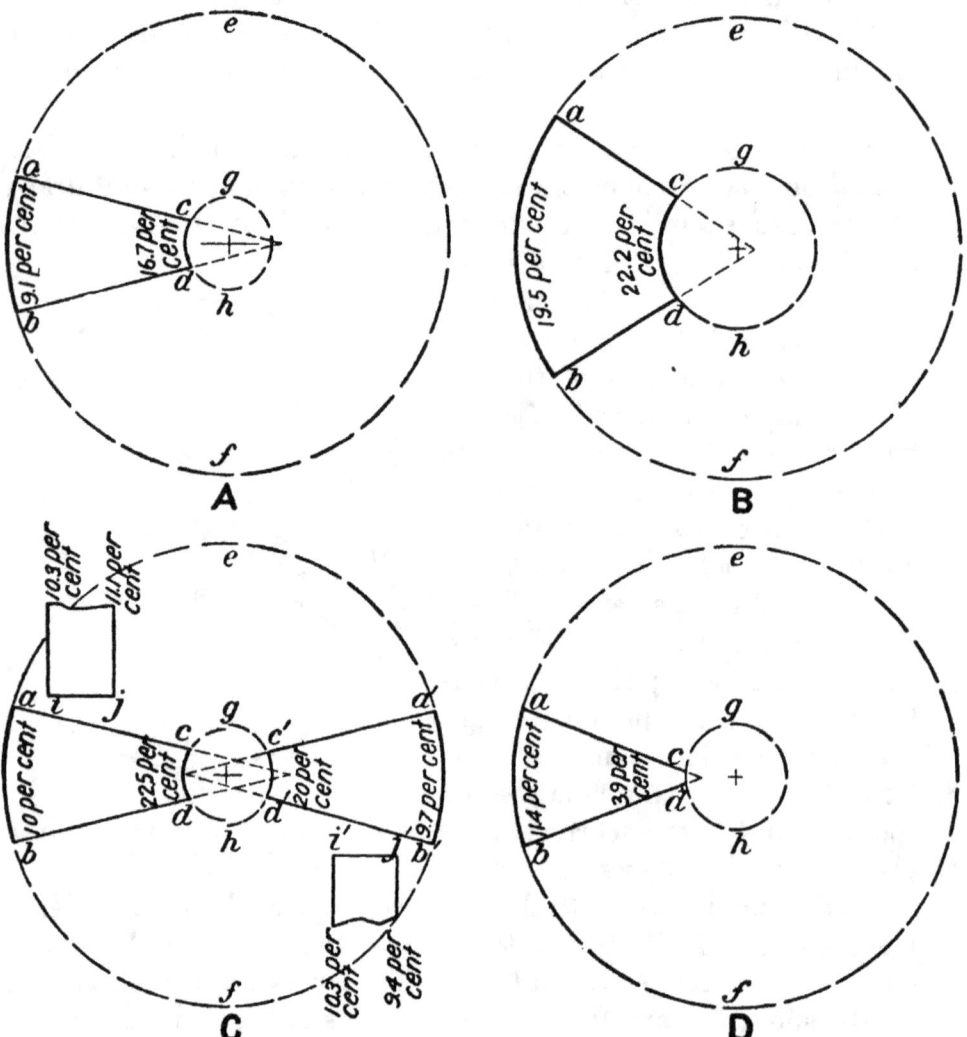

FIGURE 10.—Outlines of improperly designed Vezin sampler spouts.   A, B, and C, spouts with inside edge too wide and the continuation of the sides meeting beyond center of revolution; D, inside edge too narrow, lines meet inside of center of revolution.

that the point j is higher than the point i (fig. 10, C), the fine ore will tend to gravitate toward i, giving too small a proportion of the fine ore in the sampler under the conditions shown in A, B, and C, and too great a proportion under those in D.   If the chute is tipped so that i is higher than j the conditions, of course, would be reversed.

Views of an improperly designed Vezin machine are shown in Plate III, A, B, and C.   The contraction of the sample spout as the center of rotation is approached is evident. The measurements of this machine show that the outer extremities of the two spouts will cut a 9.8 per cent sample, whereas the inner extremities will only cut a 6.1 per cent sample. It is interesting to note that this machine was so new when photographed that it showed almost no wear. It was not found in the mills whose flow sheets accompany this paper.

Figure 11 illustrates the application of the knee-joint spout to an improperly built Vezin sampler. By adjustment of the angle of the upper part *a*, and of the lengths of parts *a* and *b*, considerable control can be exercised over the distribution of the coarse and the fine ore. The sides of the sample spout are shown tangent to the driving shaft, which was at one time the popular extent of the widening of this spout at the center. Whether fine or coarse ore preponderates in the sample would have to be determined by collecting samples at the inner and outer edges of the delivery chute. If part *b* were short enough, the coarse ore would be delivered near *d* and the finer ore nearer to *c*. With such a feed, the sample would contain too great a proportion of coarse ore in the spouts shown in A, B, and C, figure 10, and too small a proportion of coarse in a spout like D, figure 10.

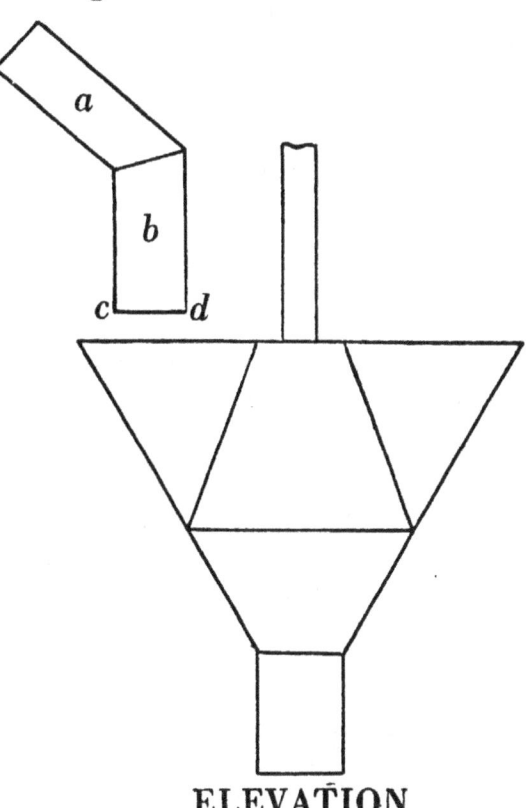

ELEVATION

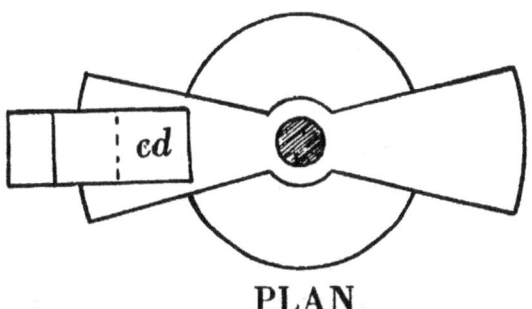

PLAN

FIGURE 11.—Sectional sketches of knee-joint chute and Vezin sampler with incorrectly designed sample spouts.

CHAS. SNYDER SAMPLER.

The Chas. Snyder sampling machine is in use in several plants which are under one management, the general system being represented by flow sheet 54 (p. 92). The principle of the machine, as shown in figure 12, is similar to that of the Vezin, in that it revolves in a horizontal plane and has spouts with radial edges. The delivery chute, however, instead of being square or rectangular as in the Vezin, is of a special shape. The bottom of the chute is annular and covers an arc of 90° directly over the sample spout, and the arc is less at the top where the ore is received from the elevators, rolls, or shaking trays. From the inside to the outside of the chute extend a large number of short iron rods that scatter and delay the falling ore and tend to break up any previous segregation. The machine has four spouts so that at all times, even if the machine should stop for any cause, one of these spouts is receiving some of the ore. What has been said of the advantages and disadvantages of the Vezin sampler also applies to this machine, except that, as it is patented and its use confined to one company, there has not been the temptation or opportunity for its misuse.

BRUNTON VIBRATING SAMPLER.

The Brunton vibrating sampler is the only machine used at two of the plants visited, and one other plant uses it in connection with the Vezin, as shown in flow sheets 2, 33, and 34.

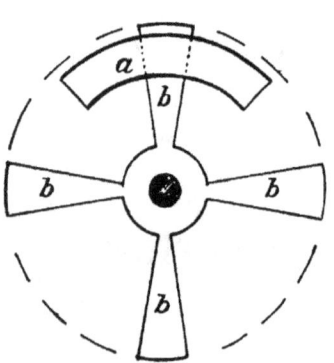

Front and side views of this machine are shown in figure 13. The ore is delivered through the chute b, which is narrowed at the lower end, as shown at c. The operating mechanism drives the diverting arm from the position a to the position a' and back again. If less than one-half is being taken for the sample, the vibrating arm rests in the position a longer than at a', thus diverting the ore for a longer

FIGURE 12.—Plan of Chas. Snyder sampler. a, delivery chute; b, sample spouts.

period of time into the reject. This device has few wearing parts, is easily cleaned, can be quickly and simply changed to cut different percentages, and with proper arrangement of the feed will take a correct sample. The correct feed, as insisted on by the inventor, is assured by delivering the ore to the machine in a thin stream parallel to the axis of vibration, which will cause the diverting arm to cut the stream across its narrowest dimension. The disadvantage of this device is that the delivery of the ore to it may be such as to give an incorrect sample. Figure 14 illustrates this possibility. The lower edge of the delivery chute is represented by $d\,h$ and the extreme right and left positions of the vibrating arm are represented by b and a, respectively. The regular movement of the arm is through a small arc, but in the figure this is exaggerated for the sake of clearness. Assume that a 20 per cent sample is being taken and that the timing is such that the arm is in the position *a* approximately four times as long as in *b*. In moving from *a* to *b* the sample spout will not be receiving ore from all points on the line *d h* for equal periods of time. Ore from part or all of the section *d e* will be falling into the sample while the arm moves from *d* to *h* and back again, from section *e f* while it moved from *e* to *h* and back again, from section *f g* while it is moving from *f* to *h* and back, and from the section *g h* while it is passing from *g* to *h* and back. If the different rates of speed at points between *a* and *b* are regarded as negligible, the sections *d e*, *e f*, and *f g* deliver to the sample, while the diverting arm is in motion, four, three, and two times as much ore, respectively, as *g h*. When the machine is, as insisted on by the in-

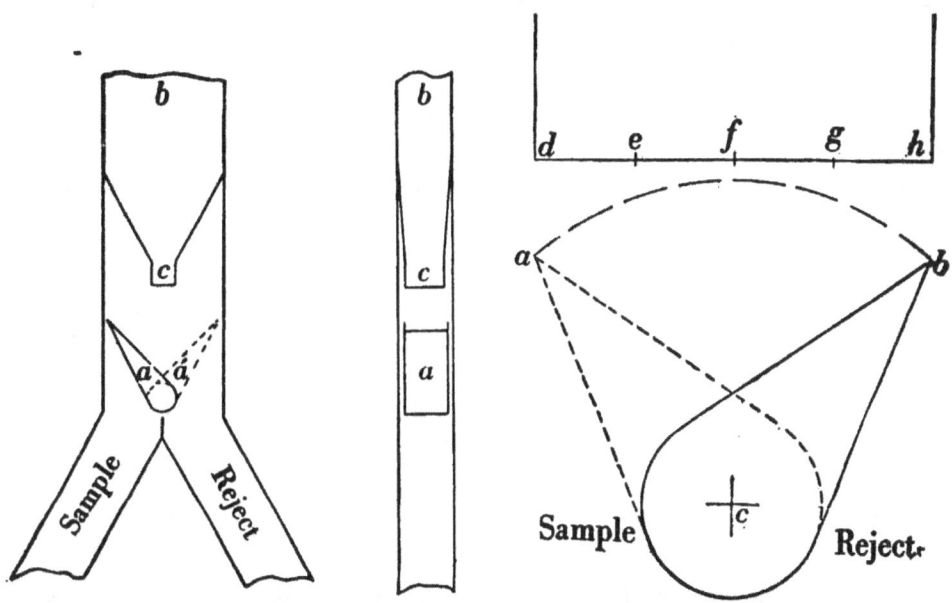

FIGURE 13.—Front and side view of Brunton vibrating sampler.

FIGURE 14.—Diagram showing path of diverting arm of Brunton vibrating sampler.

ventor, operated under a thin stream of ore, such as falls from a shaking tray, the short length of the arc cut through the ore by the vibrating arm makes these differences negligible. However, should the ore be delivered to this machine in a stream of large diameter and in a manner such as would exaggerate the natural tendency of the particles to segregate, the sample obtained will contain an incorrect proportion of either the coarse or the fine ore. Such a condition seemed to exist in two places in one plant and in one place in another plant where a narrow delivery chute was not found. This condition is illustrated in A, figure 15, in which the ore from the rolls falls into a hopper having one inclined side. At the bottom of this hopper is a short straight vertical chute that discharges directly onto the sampler. The ore in its fall first strikes the inclined side of the hopper, the fine ore sliding over to and being discharged from the opposite side of the straight chute, while the coarse ore bounds from side to side without any regularity. Following this machine is a second one operating under similar conditions but causing less segregation. The rolls of this second machine are in the position of the dotted lines shown in A. This increases the total error unless, as is the occasional custom, the sample be taken from the 20 per cent side in one machine and from the 80 per cent side in the other. B shows the conditions of delivery from an elevator, C from a knee-joint spout, and D from a screen. These require no explanation.

It is futile to attempt to prophesy positively the exact position of fine ore after it leaves an inclined surface. The physical condition of the ore partly determines this. Small samples collected by putting the hand in different parts of the stream give a rough indication of the distribution of fine and coarse ore in the stream and,

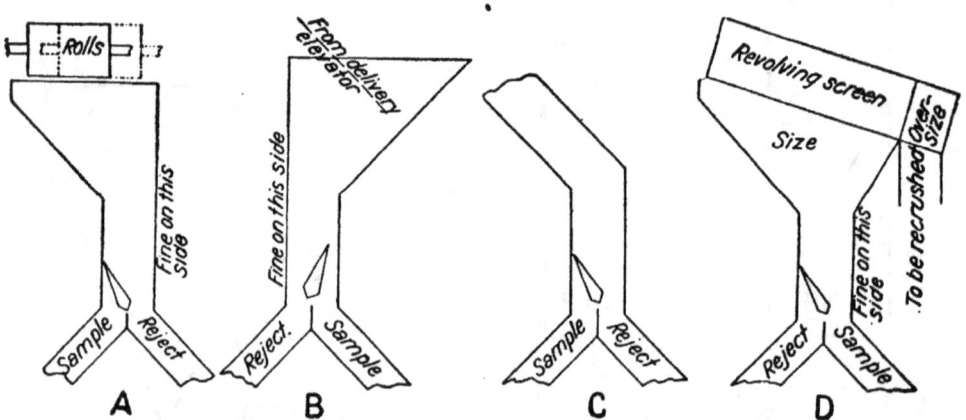

FIGURE 15.—Incorrectly arranged feeds to Brunton vibrating sampler. A, ore falls from rolls into inclined hopper; B, delivery from elevator; C, knee-joint chute; D, delivery from screen.

if more exact evidence is desired, screen or assay tests may be made. Nevertheless, the important fact remains that there is a probability of segregation in these instances that will affect the sample. As previously explained, this danger is only eliminated if a narrow spout is inserted and is kept in good repair.

BRUNTON OSCILLATING TIME SAMPLER.

The Brunton oscillating time sampler is the only machine used in six of the plants inspected. The diverting device is shown in two positions in figure 16. In construction and operation it avoids many of the dangerous features of the earlier inventions. It has a rectangular sample spout instead of one with radial sides, and as the sides of this spout are short and their cutting edges are made of special steel, there is little chance for distortion. The sampler is so balanced that it is given a rapid motion and thus takes a large number of samples, as many as 72 per minute. Through this rapid motion and on account of its oscillating in a vertical plane, it has a greater tendency to clear itself in case of obstruction or of ore which is moderately sticky. It requires little head room, so that a series of three or more machines, with the intervening rolls and shaking trays, may be erected without making a high mill. It is also quickly and easily accessible for examination, cleaning, and repairing.

Its evident disadvantage is the difficulty of handling wet and sticky ores, but this is so fundamental a fault of mechanical sampling that it is questionable as to how accurate the sampling of wet and

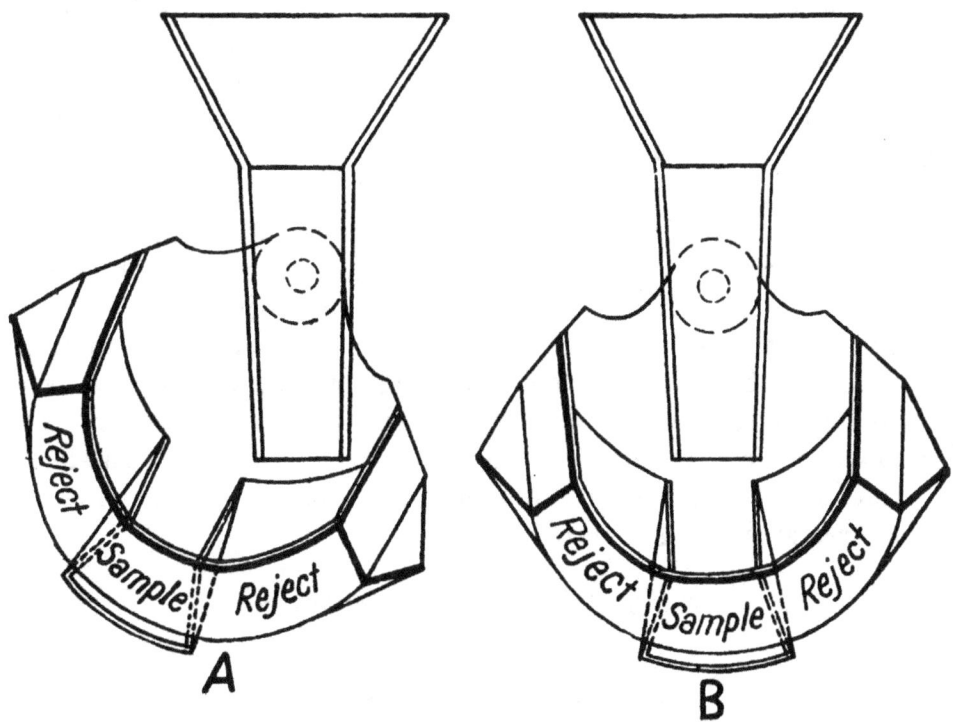

FIGURE 16.—Brunton oscillating time sampler. A, position of sample spout when ore is going to reject; B, position when ore is going to sample.

sticky ores is by any system that employs much spouting. However, the construction of this machine and the accessibility of its parts make it possible, by using plenty of manual labor, to keep the ore moving and the sample spouts clean. The principal attention required is to keep the edges of the sample spout in good repair, in order that clean cuts may be made through the stream of ore and in order to prevent pieces of ore from bounding across the partitions. In a few plants negligence was noted in this matter, several of the partitions being more than half worn away. As in all mechanical devices, the clearance between the bottom of the delivery chute and the top of the sample spout must be ample to prevent clogging.

SYNCHRONISM IN MECHANICAL SAMPLERS.

When two or more sampling machines are operated as a train, there is necessarily a constantly recurring cycle of their relative positions. For instance, if two horizontally rotating machines are so placed that one is directly beneath the other, and are given the same speed of rotation, the sample spouts will always keep the same angular distance apart. If the machines are given different speeds, the sample spout of one will catch up with and pass that of the other with a regularity like that with which the minute hand passes the hour hand of a watch. If no means of converting the sample from the first machine into a steady flow have been interposed it will strike certain points on the second with clocklike regularity. If the machines revolve at the same speed, the sample spout of the second machine may be so placed that all, a part, or none of the ore delivered by the first sampling spout will fall into the sampling spout of the second machine. If they have different speeds, this will still occur, but only during certain calculable periods. On account of the intervening rolls, chutes, and hoppers of the ordinary mill, error from this cause is remote. But instances were noted by the writer where it was probable that this synchronism had been the cause of otherwise inexplicable differences between samplings and resamplings, as well as the reason for large differences between the calculated and the actual weights of the sample. To illustrate this condition, reference is made to the following table containing data collected while the author was examining a certain plant.

Examples of synchronism in sampling machines.

FIRST COUNT.

```
S...................5 10  1  6  1  4  1  1  4  3  6  1  8  1
R...................5  1  5  1  5  1  1  7  1  5  1  5  1  5

S...................4  1  6  1  7  1  4  1  8  1
R...................1  3  1  6  1  7  1  5  1  5
```

SECOND COUNT.

```
S...................1  4  1  6  1  6  1  6  1  1  7  1  6  1
R...................1  4  1  1  7  1  5  1  1  6  1  3  1  5

S...................6  1  4  1  4  1  8  1  1  9  6  3  6  1
R...................1  7  1  5  1  5  1  1  4  5  5  1  1  7

S...................4  1  6  1  1  4  1  4  1
R...................1  7  1  3  1  1  5  1  5
```

In this table the figures in the upper row, designated "S," represent the number of consecutive samples from the first machine that passed partly or entirely into the sample spout of the second machine. The lower row of figures, designated "R," represent the number of consecutive samples from the first machine that passed entirely into the reject compartment of the second machine. The counting was continuous in each of the two examples, and a figure in the "R" row follows the figure immediately above it in the "S" row, then the next figure in the "S" row is counted. For instance, in the first example the sequence is S–5, R–5, S–10, R–1, S–1, R–5, and so on.

The value of any sample portion lies entirely in its being so fed to the following machine that a certain portion of it shall continue to the ultimate sample. If even a single sample from the first machine passes entirely into the reject of the second machine, the effect of that particular sample portion is absolutely lost, and as far as it affects the final sample it might as well have been left in the mine or have been hurled at once into the reject of the first machine. It evidently is not theoretically correct to permit as many as seven or more consecutive samples from one sampler to fall wholly into the reject of the following sampler.

On the large lots of ore handled by this particular plant this system may, by a remarkable averaging of results, give correct or at least undisputed samples. Several instances were noted at other plants where a similar condition seemed to exist, but owing to the housing and other machinery surrounding the samplers it was not feasible to get close enough to the machines to make a sufficiently accurate counting.

Contrary to the opinion of many, intervening rolls alone do not delay the ore sufficiently to entirely correct this error. Mills have been entirely reconstructed to avoid this condition. An elevator eliminates much of the trouble and the continuous delivery of the Chas. Snyder sampler lessens it, but the better system is to interpose between the samplers a hopper, a revolving barrel, or a shaking tray. Many plants have adopted one or more of these systems, partly with this object in view, and partly as incidental to an attempt to mix the ore. For instance, 8 plants use the barrel mixer, 19 use the shaking tray, and 6 use both systems, whereas 14 use neither the barrel nor the tray.

Flow sheet 19 shows a system that avoids any possibility of synchronism by collecting in a hopper the entire lot or sample after every separate crushing or sampling operation.

The barrel mixer delivers the ore in a steady stream, and, if followed immediately by a correctly built sampling machine, is a satisfactory device for this purpose. Instead of being a mixer, however, it will be found that there is a decided tendency for the coarser particles to separate from the finer and to remain nearer to the lower part of the barrel. If the barrel mixer is followed by a set of rolls, the coarser and harder particles tend to fall on a certain part of the rolls, causing uneven wear and nonuniform product.

The shaking tray has an advantage over the barrel mixer in that it combines with the delaying of the ore a feed that is steadier and of such width as may be desired. It may thus deliver the ore in a comparatively thin stream over the full width of the rolls following it, or if equipped with diverting wings will guide the ore to that part of the rolls showing the least wear, thus helping to keep them in good condition. The steady hum of the mill fully equipped with shaking trays is quite noticeable, as all the rolls receive a steady and uniform feed and are free from the jolting that results from suddenly receiving an excessive load after running several seconds or minutes on no load. Some mills have adopted the plan of placing in the shaking trays a device similar to a rake, which is occasionally inspected to prevent the passing of nuts, bolts, or even parts of a stick of dynamite that may have been left in the ore.

FINAL WORK.

The final work includes the drying, grinding, and sacking of the sample after it has reached the bucking room. At this stage of the process the size of the largest particle is planned to range from one-eighth to one-half inch in diameter and the weight of the sample from 25 to 500 pounds, the more common weight in hand-sampling mills being 200 pounds and in mechanical mills 100 pounds. When the weight is within the required limits the sample may be immediately placed on the drier, but it is frequently necessary to still further reduce the quantity at this point. At 22 of the plants inspected this was accomplished by coning and quartering methods similar to those previously described; one plant used the flat riffle commonly seen in assay offices, and in 24 plants the Jones riffle was used, in which the plane of the riffles is set at an angle of about 45° with the horizontal. This riffle is placed over a deep pan, and, as the ore falls on the riffling edges, one-half drops into the pan and the other half slides through the riffle chutes into a similar receiving pan; this process is repeated as often as necessary to reduce the sample to the desired size. Many plants are using a modification of this device called the Brunton riffle, shown in Plate III, *D* and *E*, which is similar to a double, heavily built Jones riffle. The ore is fed on horizontal dividing edges from which chutes lead alternate streams to pans placed on either side of the riffle. The importance of the work at this point is not always properly appreciated, it being too often taken for granted that a riffle will do accurate work no matter how it is built or how much it is neglected. Many riffles are made of very light material, in which case the chutes are rarely in alignment, or may have the cutting edges so bent, cracked, or even broken that it is problematical what proportions are being taken for the sample. In some of the riffles examined the chutes were entirely broken from their fastenings, permitting the fine ore to sift through to the floor or into the wrong receptacle. It was often observed that in riffling the ore was fed so fast that many of the divisions were completely clogged during a part or all of the process, and therefore these chutes were not doing their proportion of the work. Experiments have shown that it is possible to affect intentionally the sample one way or the other while riffling, even when the ore is of the fineness usual at this stage of the operation. Therefore, everything possible should be done to furnish the operator with a correct riffle and rigid instructions regarding its use.

In nine plants the Brunton one-fifth shovel or one-half shovel was used, either exclusively or in combination with other methods, for reducing the size of the sample. Figure 17 illustrates the quarter shovel. The one-half shovel is similar but has seven instead of three divisions. When this shovel is used the sample is coned or thrown into a more or less regular pile or may be left on the floor just as it falls from the delivery car or barrow. The shovel is then pushed into the pile until the compartments are as full as desired, then removed and tilted backward and downward toward the handle to allow the one-half, or four-fifths, as the case may be, to slide from the back of the shovel into the reject pile. Then the sample portion may be thrown upon a special pile. This is a rapid and fairly accurate method of cutting down the sample. However, with an ore of very high grade, the one-fifth shovel may introduce an error in the following manner: The natural way for a shoveler to work is to push the shovel radially into the pile. When the shovel is handled in this way, especially as the center of the cone is approached, the central compartment will get an undue proportion of the fine ore and the two outer compartments will get a corresponding excess of the coarse ore. It was to overcome this tendency that the one-half shovel was substituted, and with its proper use the probability of error is negligible. Owing to the size of its compartments, the Brunton shovel should not be used on lots of 50 pounds or less, and in the case of high-grade ores it is questionable whether it should be used after the sample has been reduced to 100 pounds. Nevertheless, instances were noted where samples containing pieces of ore one-half inch in diameter were fifth shoveled down to 10 pounds.

FIGURE 17.—The Brunton quarter shovel.

DRYING.

The sample, which has been reduced to 2 to 50 pounds, is dried. This is accomplished by steam plates in 5 plants, by steam coils in 36 plants, and by electric current in 4 plants.

The steam plates are iron boxes through which steam circulates, and are either installed in the open or in an inclosed room. The top or drying surface has an area of 6 to 20 square feet and is with or without raised edges. The sample is spread over this surface and left until dry.

The steam coils are built in a number of ways, the more common being an inclosed brick closet in which the shelves are a series of coils of pipe through which the steam circulates. The sample is placed in pans on the coils forming the shelves.

The temperature of steam driers is supposed to be a little above the boiling point of water for the altitude of the locality. In a few plants a record was kept of the temperature, either directly by thermometers or indirectly by the steam pressure, but in most of the plants no records seemed to be kept and the sample was, therefore, dried at whatever temperature the steam happened to be. Owing to the length of time that the sample is generally left on the drier, this lack of regularity in temperature may not be a serious matter, except with ores containing a large proportion of hydroscopic minerals, from which the percentage of water removed varies with the temperature of the drier. The importance of temperature on this class of ores will be apparent after reading the discussion regarding the moisture sample on pages 53 to 57.

Electric coils, where used, are placed in a tightly built compartment and pans containing the sample are placed over the coils on steel supports. The temperatures are indicated by thermometers, and at two plants a special device controlled the temperature automatically by shutting off the current when the desired temperature was reached.

In some plants the ore is not sufficiently dried and reaches the assayer and chemist in this condition. If the moisture sample is dried in an exactly similar manner, the only error, so far as the sale of the ore is concerned, arises from the liability of slight explosions of steam during assaying. However, the moisture sample is often dried on a separate and hotter drier and is rarely removed therefrom while it still contains water. If the assayer does not redry the pulp, and weighs it while it is slightly damp, the results of his work will be low in proportion to the amount of water contained in the sample and, the value of an ore being calculated on the dry weight as determined by the moisture sample, the seller will not be credited with the full amount due him.

The drying of easily oxidized sulphide ores is a troublesome operation with any form of drier. In some plants, although the temperature of the drier was not excessive, the sulphur eliminated from the sample made the whole room unpleasant. If this roasting adds an appreciable amount of oxygen to the sample, the assay will be correspondingly low and, should the sample lose more sulphur than it gains oxygen, the assay will be high.

FINAL GRINDING.

The sample, after being dried, is ground fine enough to pass a 40 or 60 mesh screen. Twenty-seven plants use the cone-and-ring grinder for this work, six use the disk grinder, four use the cone-and-ring and disk grinders, one uses a combination of the disk grinder and rolls, and five use all three of the grinding devices mentioned. Advantages are claimed for each system, but all are satisfactory if the devices are constructed of the proper material and are kept in good repair and do not permit the loss of dust.

There is a wide difference in the specifications of the iron for the grinding surfaces. If they are made of very hard iron they soon take a polish which causes them to grind very slowly and for that reason a medium hard iron is specified. On the other hand, if the iron is too soft, the parts will lose an appreciable amount of iron, thereby rendering the sample incorrect; for instance, if the grinder loses one-fourth of a pound of iron in grinding a 25-pound sample of ore, the results of the assay will be approximately 1 per cent low. Also a soft-iron grinding surface will take up some of the metallic gold or silver, if the sample contains any, and may in this way cause a more serious error than through the addition of iron to the sample. This danger is so well recognized by operators that it is customary to clean the grinder by putting through it barren quartz or slag in order that the enriched surface may be ground from the cones and rings. In some plants a part of the sample is used for cleaning and in one case fully one-quarter of the sample was thus used and thrown away. This practice can not be recommended. Another condition noted was the number of holes in the grinding surfaces, which were sometimes one-fourth of an inch in diameter and numbered as many as five in the same cone. It is evident that these holes should be thoroughly cleaned after each grinding, but instances were observed where this was perfunctorily done, with a probable loss from one sample and enriching of the next. A not uncommon practice is to fill these holes with Babbitt metal. This practice is inadvisable, as the softer metal may collect some of the metallics (particles of metallic gold, silver, or copper) from the ore. Several plants have adopted the plan of immediately discarding a cone or ring the moment that a small hole appears and of never using a ring or cone for grinding low-grade ore after passing through it a high-grade ore. That the iron will take up metallics has been proven

34915°—16——4

by the results of assays made upon the previously barren cleaning materials.

After final grinding, further reduction in the weight of the sample may be accomplished, if desired, by some of the methods previously described under coning and quartering or riffling. In 24 of the plants visited this reduction is to an approximate weight, or to such a quantity as seems to the operator to be sufficient for the final samples required. In 11 of the plants it is required that the sample, after final reduction, shall weigh an exact amount, usually 12 ounces. This reduction is generally accomplished by reriffling the first rejects or mixtures of the rejects and the sample until the required weight is obtained.

SCREENING.

The sample is next passed through a brass, bronze, or copper wire sieve, varying from 80 to 200 meshes to the linear inch.[a] These are the extreme sizes, the more common being either 100 or 120 mesh. In some plants the mesh used depends on the grade or character of the ore. For silver ores of medium grade it is the custom to use 80 mesh; for high-grade silver or low-grade gold ores 100 mesh; and for high-grade gold ores either 120, 150, or even 200 mesh.

In the last few years there has been a great advance in the manufacture of wire screen cloth, and it can now be obtained in guaranteed quality, both as to accuracy of the mesh and freedom from missed wires. Nevertheless, many of the plants use a cheaper and unreliable quality. Screens were noted that had several missed wires for their full diameters and others had wires crowded together, giving openings that were twice as wide as called for by the standard mesh. The better screens were so accurately made that there was no place for part of the sample to collect and enrich the next sample, but others had loose-fitting frames containing many crevices that formed lodging places for ore and, on the whole, had a generally abused and neglected appearance. Small holes in screens may be closed with a drop of solder, but the safer course is to discard the entire screen as soon as a hole appears. One hole often indicates a general weakness and, if succeeding holes are repaired with solder, there will be such a quantity that some of the metallics in the ore may be ground into it.

The resulting oversize is reground by a number of methods; 22 plants use the bucking board alone, and 13 plants regrind until all the sample passes through the screen, regrinding the oversize and

---

[a] For a discussion of the effects of variations in sieves on the results of sieving, see Stratton, S. W., Specifications for and measurements of standard sieves, Circular 39, Bureau of Standards, 1912, 14 pp.; Pearson, J. C., and Rudolph, R. J., Variations in results of sieving with standard cement sieves, Tech. Paper 29, Bureau of Standards, 1913, 16 pp.; Standardization of No. 200 cement sieves, Tech. Paper 482, Bureau of Standards, 1914, 51 pp.

using the bucking board for the last obstinate fraction. Mechanical power was used for operating the screens in 3 plants and mechanical bucking devices in 2 plants, but in all others visited both the screens and the bucking board were operated by hand.

USE OF WASHERS AND BRUSHES.

It is the general practice to place one or more iron washers in the screen with the sample to break up the lumps that form even in a dry sample. In several plants a stiff paint brush was used in addition to the washers, which resulted in an added danger of enriching a following lot or of causing loss from dust. Both the washers and the brush aid in forcing the particles through the meshes of the screen. It is not wise to overdo this forcing, as it may cause long wires of metallic gold, silver, or copper to pass through the meshes before they are properly broken up. Such samples give very erratic results when assayed and are a prolific source of trouble to the assayer.

In Plates IV and V are shown photomicrographs of two slides made from gold ore pulps from commercial samples, in which the wires and plates can be plainly seen. Everything showing black in the illustrations is gold, as the other material was removed by acids before the slides were mounted. Some of the gold was rolled into irregular cylinders as much as 0.03 inch in length, although the mesh of the sieve used had openings less than 0.004 inch. It has been calculated that, if the assayer weighed into his first crucible one more of these wires than he weighed into the second crucible, the difference in the assays reported would be over one-half ounce of gold per ton.

With ore containing metallics the only safe course is to discard the washers and take the time necessary to put the pulp through the screen by shaking. It required persistent use of washers to force through the 120-mesh screen the hook of gold showing in the upper right hand corner of Plate V, B, and the long wire near the upper left-hand corner of the same figure.

TREATMENT OF METALLICS.

It sometimes happens that material which it is impracticable to grind sufficiently fine to pass the screen remains on it. This residue may consist of pieces of steel, iron, or parts of copper detonating caps, but it often contains pieces of gold, silver, or copper belonging to the ore and frequently so malleable that they persist in the form of wires or plates. These are called "metallics" and are given special treatment. The steel or iron is removed by a horseshoe magnet and, if the operator has sufficient confidence in his judgment, he may pick out and throw away the pieces of copper cap; however, the following plan is safer and is the one generally adopted: After all the more friable material has passed through the screen, the fine ore is carefully weighed. The metallics are then carefully collected and wrapped in a piece of glazed paper which is placed in a separate sack and sent to the assayer with the sample of fine ore. The assayer then makes separate assays upon the fine ore and the metallics and from the weights of each he calculates the assay of the total sample. Instances were noted where this obstinate residue was discarded and so was not represented in the assay. This does not always work to the disadvantage of the seller as, in some instances, the residue has been proven to consist of barren material. Therefore, the only safe procedure is to take the time necessary to grind everything sufficiently fine to pass the screen or else consider the residue as metallics and treat it as described above.

MIXING THE PULP.

The finely ground sample is called the pulp. Seven of the plants visited used the so-called Anaconda mixer, which mixes the pulp in a cubical box by rotation about a horizontal axis which passes through diagonally opposite corners of the box. This apparatus is giving general satisfaction, it is easily cleaned, and does away with the personal equation. Four plants pour the pulp in a steady stream on the center of a rapidly whirling gold pan, repeating the process several times. Three plants cone and quarter the pulp, either on a glass plate or on a bucking board. One mixes the ore by turning it over several times with a spatula on the bucking board. Twenty "roll" the pulp either on oilcloth, manila paper, or on a special glazed paper. This "rolling" is an indescribable process, and must be done correctly or segregation of the particles is sure to occur.

All of these systems may prove satisfactory if the final division of the pulp into separate samples is properly performed. In other words, while the mixing may be done carelessly or under an imperfect system, this may have no effect if the proper system is adopted for the final division.

DIVIDING THE PULP SAMPLE.

Several methods of dividing and sacking the pulp sample are in use. Twelve plants cone the pulp on a bucking board, a rolling cloth, or a special rubber sheet, flatten the cone into a cake and, by means of a small rectangular piece of tin, divide the cake into as many sectors as is required. Each of these portions is then put into a paper sack. Twelve plants spread the pulp into a roughly outlined circular cake, or draw it into a long ribbonlike pile, and then fill the various pulp sacks by taking portions from the pile at random with a spatula. In both of the above systems results may be affected by the personal equation of the workman and, in case the pulp has not been previously well stirred, or if the sweepings are not carefully divided, an error is possible.

In the majority of these systems any excess of pulp is discarded, resulting in the throwing away of some of the fine material that it is difficult to pick up with a spatula or piece of tin. The extent to which this will affect a sample was not determined, but, in any event, it would be safer and have a better effect upon the workman to insist that all the pulp be sacked.

Seventeen plants use the riffle for division, utilizing all of the pulp. The riffles used are of various forms, the more common being a small one of the Brunton type. With this system all of the pulp is quickly and accurately divided, it is all used and the lack of any or of thorough mixing is immaterial.

One plant uses an ingenious mechanical device for this purpose. The pulp is fed, without previous mixing, into a rapidly revolving reservoir inclosed in a glass cylinder. Centrifugal force throws the pulp into four paper sacks attached to as many spouts leading from the reservoir.

PULP CONTAINERS.

In four plants bottles are used for holding the pulp; all the others used paper sacks. It is the general custom to place an inner sack, which is either merely folded or sealed with a paper label marked with an identifying name or number, in an outer mailing sack of heavy paper, which is usually sealed with wax. In all cases there is reserved, in a protected room, a sealed sample of every lot of ore, to be used as an umpire sample should the buyer and seller be unable to agree upon the assay results obtained from their parts of the sample. It is claimed that during its transit through the mail, or while standing on a slightly vibrating shelf of a mill, the pulp in these containers will show some concentration toward the bottom of the sack. Therefore, most assayers remix the pulp from each sack or bottle before weighing the portion for the assay charge.

TAKING THE MOISTURE SAMPLES.

As the payments for the ore, the treatment charges, and the mill and smelter operations are based on the dry weight of the ore, an accurate determination of the moisture in any particular lot is of importance.

It is customary to determine this on several samples—two to four from each carload, or an equivalent weight—and the average of these results is taken as the total moisture content. With very high-grade ores there may be as many as 10 moisture samples taken from a single car. It is interesting to note that the large variation often occurring between individual moisture samples is generally not considered as an impairment of the value of the final average.

As the water in an ore is not a fixed quantity and may vary from day to day with the state of the weather, the moisture determination is one of the most difficult problems confronting a sampling plant. Theoretically the moisture sample, if obtained from the car, should be taken at the time the ore is weighed, and it is so taken in some plants; but in many plants it is taken during the unloading of the ore, regardless of the amount of time that has elapsed since the weighing. If the ore is received in a box car the man who is to take the sample waits until the unloaders have exposed fresh surfaces in each end of the car, when, with a small scoop, he takes from 2 to 5 pounds of ore from these faces, moving the scoop from the bottom to the top of the pile and endeavoring to get a proper proportion of the fine and coarse ore. He places his sample in a pail with or without a cover and carries it to the weighing room.

In another method the sample is taken from one or more holes dug from the top to the bottom of the ore in the car. Other plants take a grab sample from the regular assay sample obtained by coning and quartering or by fractional shoveling. When ore is received in dump cars or fed to the crusher from bins, the moisture sample is taken during the unloading of the cars or while the ore is passing from the feed chute to the crusher.

In taking moisture samples, as in all forms of grab sampling, abundant opportunity is afforded for manipulation and error. The scoop is driven in and out rapidly at various points on the freshly exposed face of the pile of ore, and if the usual narrow scoop is used it soon fills and coarse ore quite frequently is forced to the top of the scoop and falls off, leaving an undue proportion of fine ore in the sample. If the ore is in a box car the top inch or two will be dryer than the rest. If there be many pieces of ore 3 inches or more in diameter, it is usually not feasible to represent them in the sample. The tendency, therefore, is for a moisture grab sample to contain too much of the fine ore, which usually has the greater proportion of moisture. Even after ore has been crushed to 1 inch it is not unusual to find that the finer particles contain two to four times as much water as the larger pieces. Even with the most careful work, individual moisture samples from one car may vary from 10 to 25 per cent when taken by the grab system. At some plants taking the moisture sample is delayed until the ore or the assay sample has been crushed to one-fourth inch or less, when the moisture sample is taken by the grab, or the coning and quartering system. One plant takes the moisture sample at regular intervals and in small parts by a mechanical device from the reject of the last mechanical sampler. From the large amount so obtained a sample of the required weight is separated by riffling.

MOISTURE SCALES.

With ordinary ores the weight taken for the moisture determination varies from 2 to 5 pounds. Special scales are used for weighing, which have the beam divided as usual into pounds and ounces, and into per cents and decimal parts of 1 per cent in the opposite direction. One form of scale has a single beam with two riders, the larger rider indicating 1 to 99 per cent and the smaller, on an extension to the main beam, 0 to 1 per cent in tenths. In a later form there are two parallel beams of equal length, one graduated from 0 to 99 per cent and the other from 0 to 1 per cent in hundredths.

WEIGHING THE WET SAMPLE.

In weighing the wet sample an error is introduced by the general custom of dumping on the scale pan a pound or two more of ore than is required. The excess is removed by hand from the top of this pile and if too much is so removed at any one time, some is allowed to slip back through the fingers into the pan until the scale balances. In taking ore from the pan, the fingers are first extended outward and downward over the pile in the pan and are then brought together to hold the ore they surround; thus the coarse part mostly remains in the hand while the fine slips through the fingers into the scale pan. This segregation is more evident near the end of the balancing, when the last handful has underloaded the pan and a small stream of fine ore is allowed to sift through the fingers until the balance is obtained. If the balancing is done by means of a small scoop, there is less danger of error at this point.

DRYING THE SAMPLE.

The moisture samples are dried in the same way as the assay samples and generally one drier suffices for both purposes. The time of drying varies but it is usually from 12 to 24 hours. As the moisture sample is ordinarily dry some time before the results of the assay sample are known, it is customary in nearly all plants to place in the drier before the end of the day shift all the moisture samples taken during the day and to allow them to remain there until the following morning. As the temperatures maintained are usually slightly above the boiling point of water, it is assumed that even in the case of an easily altered sulphide there is no danger incurred in leaving the samples in the drier as long as may be convenient; however, a hot drier and long exposure may remove an unknown percentage of the chemically combined water of certain ores.

Reference has been made previously to the effect of using a different drier and length of period of drying for the moisture and assay samples. The correct system is to use the same drier, temperature, and length of period of drying for both samples. With such precautions, it is immaterial whether or not the last traces of moisture are expelled.

MOISTURE CALCULATIONS.

The dried sample is replaced on the same scales as is used for weighing the wet sample, the loss in weight of the original sample (moisture) being read directly, with this type of scale, from the beam as percentage loss in weight. When the moisture sample is taken from the shipment directly after weighing, the percentage of moisture contained in the ore is that actually shown by the scales, but when the moisture sample is taken from the ore after crushing, some plants make a correction for evaporation in the indicated percentage of moisture. During passage through the rock breaker and rolls and while passing through elevators and chutes, most ores lose some moisture by evaporation. To this fact is principally due the custom of taking the moisture sample before or during the unloading of the ore. To correct for this drying loss, when the moisture sample is taken after the crushing, several plants add an arbitrary percentage to the percentage actually determined. The practice in this respect is not at all uniform, and a plant may add various percentages, either because of an assumed difference in the character of the ores received, or because of a requirement in the contract of the sale of an ore. At one plant the ordinary addition is 10 per cent of the percentage of moisture determined, with a maximum addition of 1 per cent of water. For example, if the indicated moisture content was 7 per cent, the assumed actual content would be 7.7 per cent, but if the moisture content as determined should be as much as 15 per cent, the actual content would be assumed to be 16 and not 16.5 per cent. With some ores this correction is applied at all times of the year, on others during the winter months only, and on still others during the summer months only. With certain ores sampled under contract it is specifically provided that no addition shall be made at any time of the year. At another plant the amount and the maximum added is applicable in all months of the year, with an agreement in all cases that no settlement shall be made on a basis of less than 1 per cent moisture.

There seem to be fewer disputes and greater satisfaction at those plants where the moisture sample is taken from the assay sample after it has been crushed to one-fourth inch or less. Nevertheless the addition of an arbitrary percentage, when the humidity may vary from zero to 100 per cent, is not a precise method and is therefore the cause of not a little of the unexpressed dissatisfaction and suspicion on the part of some sellers concerning the whole operation of ore sampling. The man who may discover or invent a scientific and commercially feasible method of determining the moisture of an ore under all conditions will render a great service to the ore-sampling industry.

FINENESS OF CRUSHING.

The maximum size to which a certain quantity of ore should be crushed before a predetermined proportion of its weight may be safely taken for a sample is not uniformly agreed upon. Some plants have posted in their mills a table showing the minimum weights of sample allowable for various sizes of ore. These weights, which are shown in the table following, are based upon extensive experiments and calculations.[a] The weights and sizes given are for ordinary gold ores and it is suggested that these weights can be greatly increased in the case of a known low-grade silver ore. Whenever obtainable, the size of the crushing at various stages has been indicated on the different flow sheets given in connection with this paper.

Smallest permissible weight of sample for varying sizes of crushing.

When crushed to—	Smallest permissible weight, pounds.
Two inches	10,000
One and one-half inches	5,000
One inch	2,000
Three-fourths of an inch	1,000
One-half inch	400
Three-eighths of an inch	300
One-fourth inch	200
Threet-sixteenths of an inch	100
One-eighth of an inch	75
Six-mesh	50
Ten-mesh	25
Eighteen-mesh	10
Thirty-mesh	4
Fifty-mesh	1

A number of plants follow this rule closely, but the larger number have no set rule as to the relation between the size of the sample and the fineness of crushing. The general intention seems to be to make the preliminary crushing sufficiently fine to fulfill the requirements for a 1,000-pound sample, following with one or more reductions until a size is obtained desirable for a 25 or even 10 pound sample. Although in many cases this is apparently the intended plan, it is frequently not consistently followed. In a large number of plants the faces of the rolls were found to be so corrugated that pieces of twice the intended diameter passed through them freely, and in many samples weighing between 10 and 50 pounds were noted pieces of ore of one-fourth to one-half inch in diameter. In certain plants the crushing seemed sufficient for lots weighing 80 tons or

---

[a] Brunton, D. W., Theory and practice of ore sampling: Trans. Am. Inst. Min. Eng., vol. 25, 1897, p. 826.

more but inadequate for the 10 or 15 ton lots that were being handled without any change in the setting of the rolls or machinery.

In 15 of the plants visited by the writer the ore is screened at different stages, either for sampling or incidental to crushing the ore for roasting. Screens increase the amount of dust, are often inaccessible, difficult to clean, and, if used after a sample has been taken from the ore, increase the danger of enriching subsequent samples. Moreover, if rolls are kept in good condition and a sufficient margin of safety is allowed in the fineness of crushing, screens are unnecessary.

As a rule, and especially is this true of custom plants, the mills seem to be planned to handle a certain average grade of ore and the crushers and rolls are consequently set for producing a product of a fixed size. While this may seem to be expensive, unless the grades of the ore are well known, it is much the safer plan. In some plants a preliminary one-fifth to one-half shovel sample is made with the silver and lower grade gold ores but omitted with higher grade or unknown ores.

In custom plants where all grades, and for the most part unknown grades, of ore are received it is the universal practice to arrange the mill for the lower grades and, when finer crushing or a larger sample is called for, one or more of the preceding sampling machines is thrown out, thus passing along to the following rolls 5 to 10 times the usual amount of ore and keeping well within the original margin of safety.

SAMPLING INCOMPLETE LOTS.

Frequently a mine or a custom sampling plant ships a lot of ore in several separate cars that reach the purchaser's sampling plant on different days. In order to keep the plant working steadily and to prevent demurrage that would accrue if the earlier received cars were held until the lot should be completed, it is customary in many plants to take from each car, as it arrives, a certain proportion to be held in the bin until all the cars in the lot have been similarly sampled. The samples so obtained from the several cars are then combined and the sampling carried forward as usual. This is called "part-lot sampling." Except in the case of ores of moderate grade, it is unsatisfactory, particularly as this preliminary sample is usually taken by fractional shoveling. If several cars are shovel sampled at the same time and by the same gang of men, it is possible that there may be some uniformity in the shoveling and the mistakes of counting may be about the same in each car, whereas, if the shoveling is done on separate days, possibly by different men and with ore which varies in grade with each car, the ordinary errors of this system will be greatly increased. When the ore is sampled in this manner only at the smelter or reduction plant, the existence of even a very large error may never be known, as the reject in every car has been bedded and its identity lost. If, however, the ore so sampled comes from a custom sampling plant, any error is quickly discovered. The inherent errors of this method, and the endless and futile disputes arising from it, have caused certain custom samplers to refuse to permit its use on any but the very lowest grade of ores shipped by them. In some plants the part lots are held in covered bins, but in other plants they are piled on the floor, against the wall, or in open stalls, for the days or weeks necessary for the completion of the whole lot. Many such piles were noted covered with dust from other piles. In view of the danger of contamination and the added danger of intentional salting, such storage is risky. However, although the part-lot question is a troublesome one, at present there seems no feasible way to eliminate it entirely. At one time the suggestion was made that this trouble could be avoided by not making lots larger than a single carload. It was also claimed by the railroads that their system of freight tariffs made the single car lot obligatory. But the inconvenience of this plan, the lessened capacity of the sampler, the extra cost of plant clean-ups, the delaying of settlements, added to the strenuous objections of the middleman, made it so unpopular that it was not generally adopted.

CLEANING THE SAMPLING PLANT.

In going through the different plants the cleanliness of some and the lack of cleanliness of others were very noticeable. In some of the plants the examination was made quickly and comfortably, but in others it was inconvenient even to trace the flow through the mill on account of the piles of ore and dust that had accumulated during several days, if not weeks, of sampling operations.

In the best cared for plants, cleaning the mill is considered one of the most important operations and is never omitted. In plants where the hand method of sampling is used the sweeping of the floors usually accompanies the sampling operation, and the only precautions necessary are to properly clean the rolls, riffles, and other machinery. However, in mechanical sampling plants, there is always the possibility that rich pieces of ore may lodge somewhere en route and enrich a succeeding lot. In the best cared for plants 15 to 30 minutes are devoted to cleaning the mill and machinery after the sampling of each lot. This is an expensive process and, if the day's business includes the sampling of many small lots, the loss of time is serious. It is worth, however, all it costs, especially from the buyer's point of view, for 10 pounds of rich gold ore will raise the value of 10 tons of ore appreciably, whereas the same quantity of barren rock would make but an insignificant change in the assays.

In some plants the cleaning is done solely with brushes and brooms; in others compressed air is used. In either case care should be taken that the bulk of the ore that has scattered over the floor or around the the rolls and other machinery is continued along the regular route before the sampling machines have been stopped; otherwise, the sample will not contain a proper proportion of these cleanings and thus be favorable or unfavorable to the buyer.

There is no uniform practice in handling the ore that accumulates at the bottom of the elevators. At some plants the elevator boots are rarely cleaned; at others the bottom of the boot is dropped after each lot and the ore removed. These cleanings may be thrown into the feed of the ore passing to the next step in the sampling process or they may be entirely rejected and thrown into a general pile which is removed at some irregular period. This is generally the disposition made of most of the ore unavoidably scattered about the crushers and rolls.

In cleaning the finishing rolls or the various grinders it is common practice to put through them barren material such as sand or slag. The hand sieves may be similarly cleaned. This cleaning should remove from the machines the particles that stick to the softer iron or collect in any of the crevices of the machine. It is very evident, however, that it should be followed by a more thorough cleaning with a brush, compressed air, or both.

Another plan, previously referred to, is the use of some of the sample itself for cleaning. With this method, a few ounces of the sample are put through the grinder and then thrown away, and the rest of the sample immediately ground with or without further cleaning of the machines. The idea seems to be that the part thrown away has removed the remnant of the preceding sample and that, at the worst, the succeeding sample will be contaminated only by that part of the ore left in the mill. Aside from the fact that the part thrown away belongs to the sample, it is possible that the buyer is unintentionally salting the ore to his own disadvantage with the more malleable and persistent metals of the previous sample.

DANGER FROM DUST.

In many sampling plants there is so much uninclosed machinery that dust is blown all over the mill and fills every nook and cranny. In other plants the wind is occasionally permitted to sweep through the mill, and this, together with long drops for the ore and careless handling, causes a cloud of dust to escape like that of a winnowing mill. The sweeping of the finely ground ore from the floor or open drier, or the riffling at various stages, is often performed very carelessly.

In a sampling or dry crushing mill the dust collecting on beams and elsewhere will generally show a slight concentration of gold and silver, increasing with the distance from, and especially above, the point of origin. On the contrary, however, weekly samplings of this dust in one mill, and over several years of operation, has shown that beyond a varying distance, this tendency is reversed and the concentration decreases at a greater rate than it increased for the shorter distances. Therefore, carelessness in the prevention and disposition of this dust does not always militate one way in the sampling of an ore.

In some plants the preliminary sample is carried in canvas sacks to the point where it receives final treatment. Owing to the difficulty of thoroughly cleaning sacks, especially if the ore is damp, this custom seems inadvisable. The meshes of the sack may contain barren rock or valuable metals, causing a robbing of one sample and a consequent enriching of another. Aside from this danger, cleaning the sacks often creates dust that may be lost or that may settle upon another lot.

DANGER FROM SALTING.

Some of the incidental causes of unintentional enrichment of samples have already been mentioned. There remains, however, a consideration of intentional enriching or salting, special precautions against which are necessary. Occasionally scattering ashes from a gold-loaded cigarette or pipe, squirting a gold solution through a knot hole, or adding a weighed amount of high-grade ore to the sample at a certain point is attempted. On the other hand, there occasionally crops up the sample man who puts barren sand or low-grade ore in a sample sieve, into a hollow bucking hammer, into a hole beneath the bucking board or into the framework of the grinder. Many pages could be written describing the schemes, both simple and elaborate, that have been used in the past for salting. Nevertheless, however unbusinesslike and improbable such trickery may seem at modern plants, safety seems to lie only in taking adequate precautions against its possible occurrence. Therefore, in many plants the sellers, or their agents, known locally as "moochers," are not allowed in any of the finishing rooms, although they are permitted to watch all the processes through glass windows from a comfortable room. In some plants similar precautions are taken at several points in the mill. It is quite usual to collect the final sample from a mechanical sampler in a locked hopper or car and to erect partitions of wire screen of coarse mesh around one or more of the mechanical samplers.

COMBINATION OF SAMPLING AND ROASTING PLANT.

In a few plants the sampling and the crushing for roasting are combined. This is illustrated in flow sheets 1 and 35 (pp. 68 and 83). These plants crush, screen, and recrush the ore to one-half or one-fourth inch before the first sample is taken. Inasmuch as the ore to be roasted must be finely crushed, combining these mills saves the extra crushing equipment and extra attendance of the sampling mill and the expense of conveying the ore from the sampler to the crushing plant. On the other hand, the extra cost of operating a separate sampling plant is not altogether wasted, as it relieves the crushing plant of part of the work. Also, several sampling plants are not necessary, as one can be used for all kinds of ores, whether going to the roasting plant or to the blast furnace.

From the point of view of accuracy in sampling, the separate plant is to be preferred, as it can be kept free from the large quantity of dust caused by finely crushing large amounts of ore, and, further, it generally has a working force trained in this special work rather than men who are apt to make the watching of the sampling incidental to obtaining low costs in the fine-crushing department. In some works, where it is necessary to have more than one sampling plant, both plants are placed under one roof, and the ore, after being sampled, is taken by belts or cars to a separate crushing plant.

COMPARISON OF ASSAYS.

The consideration of occasional assays or of selected instances of resampling is not decisive, whether the results show accuracy or inaccuracy. To obtain a fair idea of the processes of any plant, comparisons should be made for work covering a long period of time. In Tables 1, 2, and 3 are shown assays taken from the books of one large shipper of ore and covering the entire shipments to three buying plants for a period of approximately one year. It is the custom at these plants to give the seller a pulp on which he may make an assay. The buyer then reports the assays on which he is willing to make settlement, and, if the assays of the seller on the same pulps are within an agreed difference, settlement is made on the buyer's assay. If the seller's and buyer's assays differ by a sufficient amount, either the same pulps are reassayed or a reserve pulp is sent to an umpire assayer. In each case original and duplicate samples are made, but the separation into these two samples is made after the weight of the ore has been reduced to a few hundred pounds. These three buying plants use fractional shoveling, coning and quartering, and three varieties of sampling machines, but as the original and duplicate samples and the resamples are usually made after the last coning and quartering this condition has no bearing on the differences recorded in the tables. The tables explain themselves. It is noticeable that the assays of the seller are generally higher than those of the buyer, and some of the value of the comparisons may be vitiated by the tendency of the assayer to adopt, consciously or unconsciously, a procedure that favors his employer.

Table 4 shows a comparison of assays on original and duplicate samples from one seller to one buyer during a period of 15 months. It illustrates what is being done in the way of accuracy in assaying and in making check samples on small amounts of ore. In this case, as with Tables 1 to 3, the original and duplicate samples were not separated until the amount had been reduced to a few hundred pounds, and the comparisons are not a check on the earlier sampling processes. The basis of final settlement for the lots in Table 4 was the average of both the seller's and the buyer's assays on the buyer's sample.

Tables 5 and 6 are derived from Tables 1, 2, 3, and 4. Table 4 shows the number of times certain differences in the results of gold assays occurred at a custom plant. One part of this table is based on the difference between the assays of the original and duplicate as made by the buyer, and the other part is based on the difference between the highest and lowest of all four assays. Table 6 shows these differences calculated to percentages, the lowest assay being taken as the base, or 100 per cent. Table 7 is a summary of Tables 5 and 6. In the tables O indicates original samples, and D duplicate samples.

TABLE 1.—*Comparison of assays of seller and buyer on buyer's samples.*

[Gold, ounces per ton.]

Seller's assay.		Buyer's assay.		Seller's assay.		Buyer's assay.	
O.	D.	O.	D.	O.	D.	O.	D.
0.98	0.92	0.87	0.86	2.65	2.67	2.62	2.62
1.67	1.87	1.66	1.78	2.16	2.18	2.12	2.16
1.27	1.11	1.06	1.24	2.32	2.38	2.30	2.32
1.51	1.51	1.46	1.46	2.71	2.77	2.66	2.66
1.22	1.21	1.18	1.18	2.40	2.40	2.35	2.35
1.58	1.58	1.52	1.54	3.28	3.28	3.24	3.24
1.24	1.24	1.19	1.19	4.48	4.46	4.34	4.38
2.70	2.78	2.65	2.64	5.80	5.77	5.72	5.77

TABLE 2.—*Comparison of assays of seller and buyer on buyer's assays.*

[Gold, ounces per ton. Braces indicate resamples.]

Seller's assay.		Buyer's assay.		Seller's assay.		Buyer's assay.		Seller's assay.		Buyer's assay.	
O.	D.	O.	D.	O.	D.	O.	D.	O.	D.	O.	D.
1.26	1.24	1.26	1.16	1.14	1.10	1.16	1.16	2.44	2.48	2.32	2.40
1.54	1.74	1.52	1.66	1.59	1.63	1.72	1.80	2.16	2.14	2.07	2.02
1.22	1.11	1.16	1.06	1.02	1.10	1.00	1.04	2.82	2.92	2.70	2.84
1.25	1.28	1.18	1.25	{1.61	1.75	1.58	1.71	{3.98	3.98	3.63	3.82
1.50	1.48	1.38	1.38	{1.82	1.74	1.64	1.66	{3.28	3.66	3.23	3.66
1.41	1.43	1.40	1.38	1.35	1.39	1.28	1.34	{3.09	3.11	3.06	3.04
1.23	1.17	1.18	1.08	1.04	1.00	1.02	.94	{3.26	3.24	3.18	3.12
1.07	1.13	1.04	1.08	1.50	1.48	1.45	1.36	3.47	3.69	3.42	3.56
1.42	1.40	1.40	1.36	2.98	3.00	2.92	2.93	3.38	2.96	3.26	2.80
1.96	1.94	1.96	1.93	2.10	2.10	2.04	2.08	3.08	2.92	3.00	2.86
1.91	1.87	1.94	1.90	2.56	2.73	2.48	2.68	3.45	3.37	3.40	3.32
1.19	1.18	1.18	1.14	2.09	2.11	2.06	2.06	3.20	3.18	3.14	3.10
1.15	1.25	1.12	1.14	{2.76	3.34	2.72	3.30	3.61	3.76	3.52	3.68
1.36	1.36	1.25	1.35	{2.73	2.77	2.70	2.68	3.94	3.88	3.80	3.88
1.63	1.69	1.60	1.60	2.51	2.49	2.40	2.44	3.14	3.07	3.10	3.10
1.74	1.70	1.68	1.66	2.12	2.18	2.00	2.10	3.89	4.08	3.74	4.00
1.37	1.15	1.30	1.16	2.78	2.92	2.72	2.90	3.34	3.20	3.20	3.12
1.07	1.09	1.00	1.04	2.56	2.44	2.47	2.40	3.56	3.36	3.19	3.27
1.34	1.38	1.29	1.33	2.44	2.48	2.40	2.39	3.15	3.14	2.96	3.00
1.18	1.17	1.15	1.12	2.86	2.82	2.80	2.74	4.26	4.30	4.24	4.22
1.52	1.66	1.50	1.61	2.98	2.94	2.87	2.88	4.16	4.10	3.96	3.94
1.10	1.10	1.00	1.04	2.40	2.40	2.35	2.33	4.24	4.14	4.08	4.08
.92	.98	.88	.94	3.00	2.90	2.92	2.80	4.60	4.68	4.50	4.69
1.03	1.09	.99	1.02	2.62	2.68	2.56	2.86	5.46	5.42	5.27	5.53
1.37	1.36	1.30	1.33	2.88	2.92	2.79	2.81	5.76	5.90	5.76	5.74

TABLE 3.—*Comparison of assays of seller and buyer on buyer's samples.*

[Gold, ounces per ton. Braces indicate resamples.]

Seller's assay.		Buyer's assay.		Seller's assay.		Buyer's assay.		Seller's assay.		Buyer's assay.	
O	D	O	D	O	D	O	D	O	D	O	D
{0.96	1.26	0.90	1.26	1.12	1.15	1.00	1.08	2.44	2.44	2.46	2.32
1.06	1.14	1.04	1.14	{1.48	1.10	1.44	1.04	{2.28	2.60	2.20	2.62
1.18	1.18	1.10	1.12	{1.24	1.26	1.30	1.30	{2.39	2.40	2.40	2.46
1.04	1.02	1.16	.96	1.35	1.41	1.30	1.36	2.80	2.72	2.64	2.64
.86	.74	.84	.68	1.51	1.71	1.48	1.62	{3.02	2.58	3.04	2.60
{1.32	1.50	1.32	1.50	1.82	1.86	1.84	1.76	{2.80	2.18	2.76	2.24
1.24	1.58	1.28	1.60	2.11	2.10	2.02	2.04	{3.33	2.71	3.32	2.76
{1.05	1.27	1.08	1.28	2.48	2.44	2.42	2.38	3.52	3.22	3.47	3.18
1.40	1.33	1.38	1.40	2.13	2.16	2.10	2.10	3.11	3.13	3.04	3.04
1.25	1.25	1.20	1.20	2.45	2.55	2.40	2.52	3.08	3.34	3.04	3.28
1.29	1.33	1.26	1.24	2.04	2.32	2.00	2.26	2.79	3.03	2.84	2.88
1.54	1.58	1.54	1.40	2.76	2.88	2.68	2.82	5.34	5.40	5.26	5.34

TABLE 4.—*Comparison of assays of seller and buyer on buyer's samples.*

[Gold, ounces per ton. Braces indicate resamples.]

Seller's assay.		Buyer's assay.		Seller's assay.		Buyer's assay.		Seller's assay.		Buyer's assay.		Seller's assay.		Buyer's assay.	
O	D	O	D	O	D	O	D	O	D	O	D	O	D	O	D
0.69	0.66	0.68	0.64	0.81	0.81	0.83	0.82	1.03	1.04	1.04	1.04	1.73	1.75	1.78	1.77
.78	.76	.79	.80	.59	.60	.62	.61	1.82	1.78	1.72	1.76	2.56	2.46	2.55	2.53
.60	.53	.55	.52	.54	.54	.50	.52	1.53	1.49	1.50	1.49	2.35	2.32	2.38	2.40
.54	.55	.57	.54	.68	.73	.71	.75	1.18	1.20	1.19	1.17	2.02	2.03	1.98	2.01
.67	.65	.70	.69	.71	.72	.73	.74	1.59	1.60	1.64	1.70	2.18	2.04	2.18	2.08
.70	.68	.69	.68	.70	.68	.71	.67	1.22	1.24	1.30	1.29	2.47	2.48	2.52	2.51
.36	.36	.35	.35	.68	.69	.74	.72	1.51	1.53	1.54	1.55	2.27	2.28	2.28	2.28
.49	.48	.49	.48	.84	.84	.84	.87	1.43	1.45	1.46	1.45	2.90	2.74	2.86	2.88
.94	.92	.96	.94	.52	.51	.54	.53	1.56	1.52	1.61	1.55	2.16	2.21	2.16	2.23
.86	.90	.87	.87	.88	.86	.86	.84	1.17	1.22	1.19	1.24	2.15	2.15	2.03	1.97
.31	.32	.30	.31	.75	.74	.75	.76	1.67	1.68	1.70	1.70	2.62	2.57	2.62	2.59
.58	.54	.56	.56	.58	.58	.57	.55	1.28	1.28	1.29	1.28	2.63	2.67	2.61	2.65
.94	.91	.96	.96	.86	.84	.86	.85	1.43	1.49	1.44	1.42	2.60	2.51	2.53	2.50
.50	.50	.49	.49	.46	.49	.47	.47	1.27	1.28	1.33	1.32	2.06	2.07	2.03	2.04
.98	.97	.99	.98	.67	.67	.64	.67	1.03	1.06	.98	.98	2.17	2.24	2.16	2.18
.58	.56	.57	.58	.81	.74	.80	.80	1.15	1.16	1.11	1.11	2.62	2.61	2.64	2.64
.44	.45	.42	.43	.76	.77	.75	.76	1.00	1.00	1.01	1.04	3.48	3.42	3.41	3.44
.53	.52	.57	.49	.62	.63	.62	.62	1.25	1.20	1.28	1.26	3.57	3.65	3.55	3.68
.83	.82	.86	.86	.50	.49	.52	.52	1.22	1.21	1.11	1.13	{3.60	3.62	3.44	3.42
.78	.76	.79	.79	.45	.46	.47	.47	1.06	1.05	1.00	.97	{3.33	3.31	3.34	3.30
.44	.44	.44	.44	1.05	1.06	.98	1.02	1.00	1.02	1.00	1.00	{3.10	3.02	2.89	2.91
.88	.90	.91	.89	1.66	1.66	1.56	1.56	1.17	1.16	1.16	1.16	{3.05	2.95	2.93	2.83
.74	.75	.74	.72	1.87	1.81	1.92	1.89	1.71	1.74	1.75	1.79	4.53	4.56	4.55	4.55
.79	.80	.81	.82	1.05	1.06	1.08	1.06	1.73	1.63	1.74	1.72	3.77	3.72	3.82	3.86
.74	.76	.71	.72	1.63	1.64	1.63	1.63	1.88	1.89	1.88	1.87	4.02	3.91	3.96	3.91

TABLE 5.—*Averages of results in Tables 1, 2, 3, and 4, showing the number of times differences of various amounts occur between the highest and lowest assay of a sample of ore.*

Difference between highest and lowest assay, ounces of gold.	Buyer's assays. From Table—				All assays. From Table—				Difference between highest and lowest assay, ounces of gold.	Buyer's assays. From Table—				All assays. From Table—			
	1	2	3	4	1	2	3	4		1	2	3	4	1	2	3	4
0.00	8	6	5	26	....	....	....	1	0.23	....	....	1	....	....	....	2	....
.01	2	2	....	27	....	....	....	9	.24	....	....	1	....	....	1	1	....
.02	2	12	4	18	....	....	....	13	.25	....	....	....	....	....	1	....	....
.03	....	4	....	3	....	1	....	20	.26	....	2	1	....	....	1	....	....
.04	1	12	2	9	2	....	....	13	.27	....	....	....	....	....	1	....	....
.05	1	1	....	1	4	3	1	10	.28	....	....	....	....	....	1	....	....
.06	....	4	2	3	2	4	2	3	.29	....	....	1	....	....	....	....	....
.07	....	2	....	1	6	6	3	4	.30	....	1	....	....	....	1	1	....
.08	....	6	3	1	2	3	1	10	.31	....	....	....	....	....	....	....	....
.09	....	1	....	....	5	5	2	1	.32	....	....	2	....	....	....	1	....
.10	....	5	1	2	8	8	3	4	.34	....	....	....	....	....	1	1	....
.11	....	2	....	....	3	3	1	4	.35	....	....	....	....	....	1	....	....
.12	1	1	1	....	5	5	....	1	.36	....	....	1	....	....	....	2	....
.13	....	1	....	1	3	3	....	1	.37	....	....	1	....	....	1	....	....
.14	....	6	4	....	4	4	1	2	.40	....	....	1	....	....	....	1	....
.15	....	....	....	....	1	1	2	....	.42	....	....	1	....	....	....	....	....
.16	....	....	1	....	5	5	2	1	.43	....	1	....	....	....	1	....	....
.17	....	....	....	....	1	1	....	....	.44	....	....	1	....	....	....	1	....
.18	1	1	1	....	3	3	3	1	.46	....	1	....	....	....	....	1	....
.19	....	2	....	....	2	2	....	....	.56	....	1	....	....	....	....	....	....
.20	....	1	2	....	1	1	2	1	.58	....	1	....	---	....	1	....	....
.21	....	....	....	1	2	2	....	1	.62	....	....	....	....	....	1	2	.\.
.22	....	....	....	....	3	3	....	....									

TABLE 6.—*Averages of results in Tables 1, 2, 3, and 4, showing the number of times differences of various percentages occur between the highest and lowest assays of a sample of ore.*

Difference between highest and lowest assays, per cent.	Number of times difference occurs.								Difference between highest and lowest assays, per cent.	Number of times difference occurs.							
	Buyer's assays.				All assays.					Buyer's assays.				All assays.			
	From Table—				From Table—					From Table—				From Table—			
	1	2	3	4	1	2	3	4		1	2	3	4	1	2	3	4
0	7	6	5	26			1	0	15					1	2	2	1
1	5	20	6	39	2	1		14	16		1					1	1
2	2	10	2	17	2	5	1	15	17			1					
3		6		11	2	5	4	16	18			2			1		
4		7		4	5	14	2	13	19			1				1	1
5		7	4	1	1	9	2	9	20			1			1		
6		2	1	2	1	3	4	11	21		1	1				1	
7	2	3				6	3	9	22			1				1	
8		4	2			5	1	4	23							1	1
9		5	3			8	1	4	25			1					
10			1			4	3	2	26							2	
11						6			28							1	
12					1	2			38		1						
13		3	1				1		40		1					1	
14			1		1	2	1		41							1	

TABLE 7.—*Summary of results in Tables 5 and 6, showing the range of differences in the samples at the different plants.*

Results from—	Difference in highest and lowest assays.			
	Ounces.		Per cent.	
	Buyer's assays.	All assays.	Buyer's assays.	All assays.
Table 1	0.02 to 0.03	0.08 to 0.09	1 to 2	5 to 6
Table 2	.08 to .09	.15 to .16	3 to 4	7 to 8
Table 3	.15 to .16	.14 to .23	9 to 10	12 to 13
Table 4	.01 to .02	.05 to .06	1 to 2	4 to 5

A study of these tables will show how widely different plants may vary in the sampling of an ore from a single mine. It will also show that although satisfactory sampling is being done in most plants, nevertheless sales of ore are being made on the basis of sampling and of differences in assays that might have been considered grotesque even many years ago.

It would be interesting to have the details of the sampling of individual lots of ore at the custom samplers and the sampling of the same lots at the smelters and mills, and for long periods of time. As said before, to be of value, the comparison must be made of all lots and not of selected ones; of individual lots and not of averages.

FLOW SHEETS OF PLANTS INVESTIGATED.

In the following pages is presented 55 flow sheets representing the sampling systems used at the 48 plants investigated. As was previously stated, some plants use more than one system and in such cases a flow sheet representing each system was prepared. The course of the sample in the finishing room is given as a separate flow sheet in each case.

DESCRIPTION OF FLOW SHEETS.

The general form of the flow sheets is a modification of that used by Brunton.[a]

The size of elevator buckets and other facts not pertinent to this paper are omitted.

Each line represents a distinct operation in the mill, the ore passing from the operation shown in one line to that shown in the line below it; where the flow of the ore is divided the different streams take the courses indicated by the arrows.

The figures after "Crusher" show the opening and width of jaws, and the proposed fineness of crushing.

The figures after "Rolls" show the diameter and width of each roll and the proposed fineness of crushing.

The percentages after "Sampling machine" indicate the approximate percentage taken by the machine for the sample; the number of pounds indicate the approximate weight of the sample at that point, based on an original lot weighing 100,000 pounds, unless a different quantity is noted near the top of the sheet.

The word "Discard," unless followed by one or more operations, indicates the elimination from further consideration of this part of the ore at that point.

---

a Brunton, D. W., Modern practice in ore sampling: Trans. Am. Inst. Min. Eng., vol. 40, 1909, p. 586.

SAMPLING-MILL FLOW SHEETS.

FLOW SHEET 1.

Railroad cars over hopper.
Unloaded into hopper.
Crusher, to 1½-inch.
Elevator No. 1.
Rolls No. 1, 16 by 36 inches, to 1-inch.
Rolls No. 2, 16 by 36 inches, to ½-inch.
Elevator No. 2.
Sampler No. 1, Brunton oscillating.

Sample, 5 per cent, 5,000 pounds.←——→Discard.
Rolls No. 3, 12 by 20 inches, to ½-inch.   Revolving screen, ½-inch holes.
Sampler No. 2, Brunton oscillating.

                 Oversize.←——→Size, to conveyor belt.
                 Rolls No. 4, 14 by 30 inches, to ½-inch.

Sample, 20 per cent, 1,000 pounds.←——→Discard. Elevator No. 3.
Sample bin.   Revolving screen, ½-inch holes.
Coned and quartered.

                 Oversize.←——→Size, to conveyor belt.
                 Elevator No. 4.
                 Returns to rolls No. 4.

Sample, about 200 pounds.←——→Discard.——————→To conveyor belt.
Put in canvas sacks.   Storage bins.
Taken to finishing room.
See flow sheet 3 for further treatment of sample.

FLOW SHEET 2.

Hopper beside railroad tracks.
Elevator No. 1.
Revolving mixer, 100,000 pounds.
Sampler No. 1, Vezin.

Sample, 20 per cent, 20,000 pounds.←——→Discard.
Hopper.
Wheeled in buggy.
Crusher, to 1-inch.
Elevator No. 2.
Barrel mixer.
Sampler No. 2, Vezin.

Sample, 20 per cent, 4,000 pounds.←——→Discard.
Shaking tray.
Rolls No. 1, 14 by 28 inches, to ⅜-inch.
Hopper.
Sampler No. 3, Brunton vibrating.

Sample, 20 per cent, 800 pounds.←——→Discard.
Rolls No. 2, 12 by 20 inches, to ¼-inch.
Hopper.
Sampler No. 4, Brunton vibrating.

Sample, 20 per cent, 160 pounds.←——→Discard.
Small hopper.
Wheeled to quartering room.
Coned and quartered.

Sample, about 200 pounds.←——→Discard.
Put in canvas sacks.
See flow sheet 3 for further treatment of sample.

FLOW SHEET 3.

Rolls, 12 by 20 inches, set closely.
Hand screen, 7-mesh.

Size.←————————→Oversize.
                   Returns to rolls.
Riffle, Brunton type; original and duplicate samples separated here.

Sample, 10 to 12 pounds←————————→Discard.
Steam drier, short time only.
Englebach grinder.
Riffle, Brunton type._____

Sample, about 6 pounds.←————————→Discard.
Steam drier.
Rolling cloth.
Riffle, Brunton type._____

Sample, exactly 20 ounces.←————————→Discard.
Hand sieve, 120-mesh.

Size.←————————→Oversize.
Rolling cloth.   Bucking board.
                   Returns to hand sieve.
Coned and quartered on glass plate.
Each quarter separately sacked.

FLOW SHEET 4.

Railroad tracks over bins.
Crusher, 13 by 24 inches, to 2-inch.
Conveyor belt.
Hopper.
Shaking tray.
Rolls No. 1, 14 by 42 inches, to 1-inch.
Shaking tray.
Elevator No. 1.
Inclined chute, 100,000 pounds.
Sampler No. 1, Vezin._____

Sample, 25 per cent, 25,000 pounds.←——→Discard.
Shaking tray.
Rolls No. 2, 14 by 30 inches, to $\frac{1}{2}$-inch.
Chute.
Sampler No. 2, Vezin._____

Sample, 20 per cent, 5,000 pounds.←——→Discard.
Shaking tray.
Rolls No. 3, 14 by 20 inches, to $\frac{1}{4}$-inch.
Inclined chute.
Sampler No. 3, Vezin._____

Sample, 20 per cent, 1,000 pounds.←——→Discard.
Shaking tray.
Rolls No. 4, 12 by 20 inches, to $\frac{1}{8}$-inch.
Sample hopper.
Wheeled in buggies to
Iron floor.
Coned and quartered._____

Sample, about 200 pounds.←————————→Discard.
See flow sheet 5 for further treatment of sample.

FLOW SHEET 5.

Sample shoveled into pan.
Riffle._____

Sample, 20 to 25 pounds.←————————→Discard.
Steam drier.   Stored for resample.
Grinder, vertical-disk pattern, about $\frac{1}{40}$-inch.
Riffle._____

Sample, exactly 20 ounces.←————————→Discard.
Hand sieve, 120 to 150 mesh.

Size.←————————→Oversize.
Rolling cloth.   Grinder, disk type.
                 Returns to hand sieve.
Riffle.
All pulp put in four sacks.

FLOW SHEET 6.

Railroad cars on trestle.
Hoppers, 500-ton capacity.
Crusher, to 1½-inch.
Elevator No. 1.
Small hopper.
Grizzly shaker.

Size.←————→Oversize.
             Rolls No. 1.
             Impact screen, ½-inch mesh.

           Size.←————————————→Oversize.
Elevator No. 2.   Returns to elevator No. 1.
Sampler No. 1, Vezin._____

Sample, 20 per cent, 20,000 pounds.←————→Discard.
Barrel mixer.
Sampler No. 2, Vezin._____

Sample, 20 per cent, 4,000 pounds.←————→Discard.
Rolls No. 2, 12 by 20 inches, to ½-inch.
Sampler No. 3, Vezin._____

Sample, 20 per cent, 800 pounds.←————→Discard.
Wheelbarrow.
See flow sheet 7 for further treatment of sample.

FLOW SHEET 7.

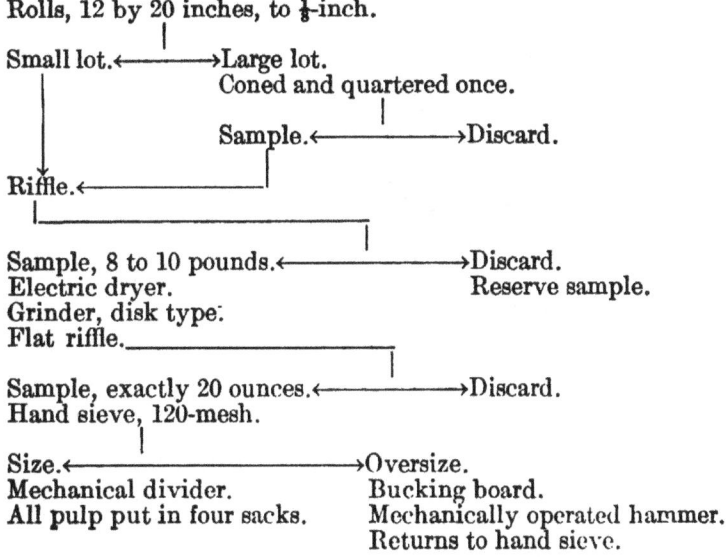

Rolls, 12 by 20 inches, to ⅛-inch.

Small lot.←——————→Large lot.
Coned and quartered once.

Sample.←——————→Discard.

Riffle.←

Sample, 8 to 10 pounds.←——————→Discard.
Electric dryer.   Reserve sample.
Grinder, disk type.
Flat riffle.

Sample, exactly 20 ounces.←——————→Discard.
Hand sieve, 120-mesh.

Size.←——————————————→Oversize.
Mechanical divider.   Bucking board.
All pulp put in four sacks. Mechanically operated hammer.
                             Returns to hand sieve.

FLOW SHEET 8.

Railroad track beside bins.
Storage bins.
Tramcar.
Crusher, 15 by 30 inches, to 1½-inch, 100,000 pounds.
Elevator No. 1.
Rolls No. 1, 16 by 36 inches, to 1-inch.
Elevator No. 2.
Sampler No. 1, Vezin.

Sample, 20 per cent, 20,000 pounds.←——→Discard.
Shaking tray.
Rolls No. 2, 14 by 27 inches, to ½-inch.
Sampler No. 2, Vezin.

Sample, 20 per cent, 4,000 pounds.←——→Discard.
Shaking tray.
Rolls No. 3, 14 by 27 inches, to ⅜-inch.
Sampler No. 3, Vezin.

Sample, 20 per cent, 800 pounds.←——→Discard.
Conveyor belt.
Closed room.
Rolls No. 4, 4 by 18 inches, to ¼-inch.
Sampler No. 4, Vezin.

Sample, 20 per cent, 160 pounds.←——→Discard.
See flow sheet 9 for further treatment of sample.

FLOW SHEET 9.

Riffle No. 1._____

Sample, about 26 pounds.←——————→Discard.
Steam drier.
Grinder No. 1, disk type, about 30-mesh.
Riffle No. 2._____

Sample, about 25 ounces.←——————→Discard.
Hand sieve No. 1, 120 or 150 mesh.

Size.←——————————→Oversize.
Rolling cloth.   Grinder No. 2, disk type.
Flattened into round cake.   Sieve No. 1, again.

Put in four sacks with spatula. Size.←————→Oversize.
Excess pulp discarded.   Bucking board.
                                          Returns to sieve No. 1.

FLOW SHEET 10.

Railroad tracks at side of mill.
Shoveled to pan conveyor.
Hopper, 100,000 pounds.
Grizzly.

Size.←————→Oversize.
Shaking tray.←——Crusher, 15 by 24 inches, to 1½-inch.
Rolls No. 1, 16 by 36 inches.
Elevator.
Revolving screen, ¼-inch holes.

Size.←————→Oversize.
Hopper.   Rolls No. 2, 16 by 36 inches.
Round spout.   Returns to elevator.
Sampler No. 1, Vezin._____

Sample, 20 per cent, 20,000 pounds.←——————→Discard.
Barrel mixer.
Sampler No. 2, Vezin._____

Sample, 20 per cent, 4,000 pounds.←——————→Discard.
Barrow.
Lift.
Compound riffle No. 1, O and D samples started here.

Sample, 12½ per cent, 500 pounds.←——————→Discard.
                                          Held for reserve
Rolls No. 3, to about ⅛-inch.
Barrow.
Lift.
Riffle._____

Sample, 50 per cent, 250 pounds.←——————→Discard.
To grinding room.
See flow sheet 13 for further treatment of sample.

FLOW SHEET 11.

Concentrates and smelters product.
Railroad cars over beds.
Fractional shoveling._____

Sample, 20 per cent, 20,000 pounds.←——→Discard.
                              On beds.

Piled near car door.
Fractional shoveling._____

Sample, 50 per cent, 10,000 pounds.←——→Discard.
                              On beds.

Piled near car door.
Barrows.
Wheeled to quartering room.
Fractional shoveling._____

Sample, 33½ per cent, 3,333 pounds.←——→Discard.
                              Reserve sample.

Coned and quartered._____

Sample, about 250 pounds.←————→Discard.
To grinding room.
See flow sheet 13 for further treatment of sample.

FLOW SHEET 12.

Railroad cars in ore house, 100,000 pounds.
Fractional shoveling._____

Sample, 10 or 20 per cent, 10,000 or 20,000 pounds.←——→Discard.
                                           Wheeled to beds.
Barrows.
Crusher, 7 by 14 inches, to 1½-inch.
Elevator.
Revolving screen.

Size.←————→Oversize.
Hopper.   Rolls, 15 by 27 inches, to about ½-inch.
              Returns to elevator.
Sampler, Vezin._____

Sample, 20 per cent, 2,000 or 4,000 pounds.←————→Discard.
Hopper.
Compound riffle.   Original and duplicate samples started here.

Sample, 12½ per cent, 250 or 500 pounds.←————→Discard.
Barrow.
To grinding room.
See flow sheet 13 for further treatment of sample.

FLOW SHEET 13.

Grinding room.
Drier, steam-plate type.
Partly dried.
Swept into barrow.
Grinder, Englebach type, to about 10-mesh.
Riffle No. 1._____

Sample, about 20 pounds.←————————→Discard.
Through funnels into buckets.
To bucking room.
Steam drier.
Grinder, Englebach type, to 30 or 40 mesh.
Riffle No. 2._____

Sample, 50 per cent.←————————→Discard.
Riffle No. 3._____

Sample, 50 per cent.←————————→Discard.
Riffle No. 4._____

Sample, exactly 20 ounces.←————————→Discard.
Hand sieve, 80 to 150 mesh.

Size.←————→Oversize.
              Bucking board or disk grinder.
              Returns to hand sieve.
Mixed and coned on bucking board.
Flattened into cake by sieve bottom.
Marked in six sectors.
All pulp put in six or three sacks.

FLOW SHEET 14.

Railroad cars to beds, where one-tenth is separated by shoveling.
              or
Railroad to crushing plant (see flow sheet 16).
Cars containing one-tenth set beside sampling mill.
Crusher, 10 by 20 inches, to 1½-inch.
Inclined chute.
Sampler No. 1, Vezin._____

Sample, 20 per cent, 2,000 pounds.←————→Discard.
Inclined chute.
Rolls No. 1, 16 by 40 inches, to ¾-inch.
Chute.
Elevator.
Impact screen, variable mesh

Size.←————→Oversize.

              Rolls No. 2, 14 by 27 inches, ¾ to ½ inch.
              Returns to elevator.

Sample, 20 per cent, 400 pounds.←————————→Discard.
Barrow.
Wheeled to quartering room.
If necessary, entire sample is dried.
Coned and quartered._____

Sample, about 400 pounds.←————————→Discard.
                                      Held as reserve.

See flow sheet 18 for further treatment of sample.

FLOW SHEET 15.

Railroad cars near beds where ore is fractional shoveled.
or
From crushing plant (see flow sheet 16).
Cars containing one-tenth placed beside mill, 10,000 pounds.
Crusher, 13 by 24 inches, to 1½-inch.
Rolls No. 1, 16 by 40 inches, to ¾-inch.
Elevator No. 1.
Impact screen, from 1 to ¾ inch.

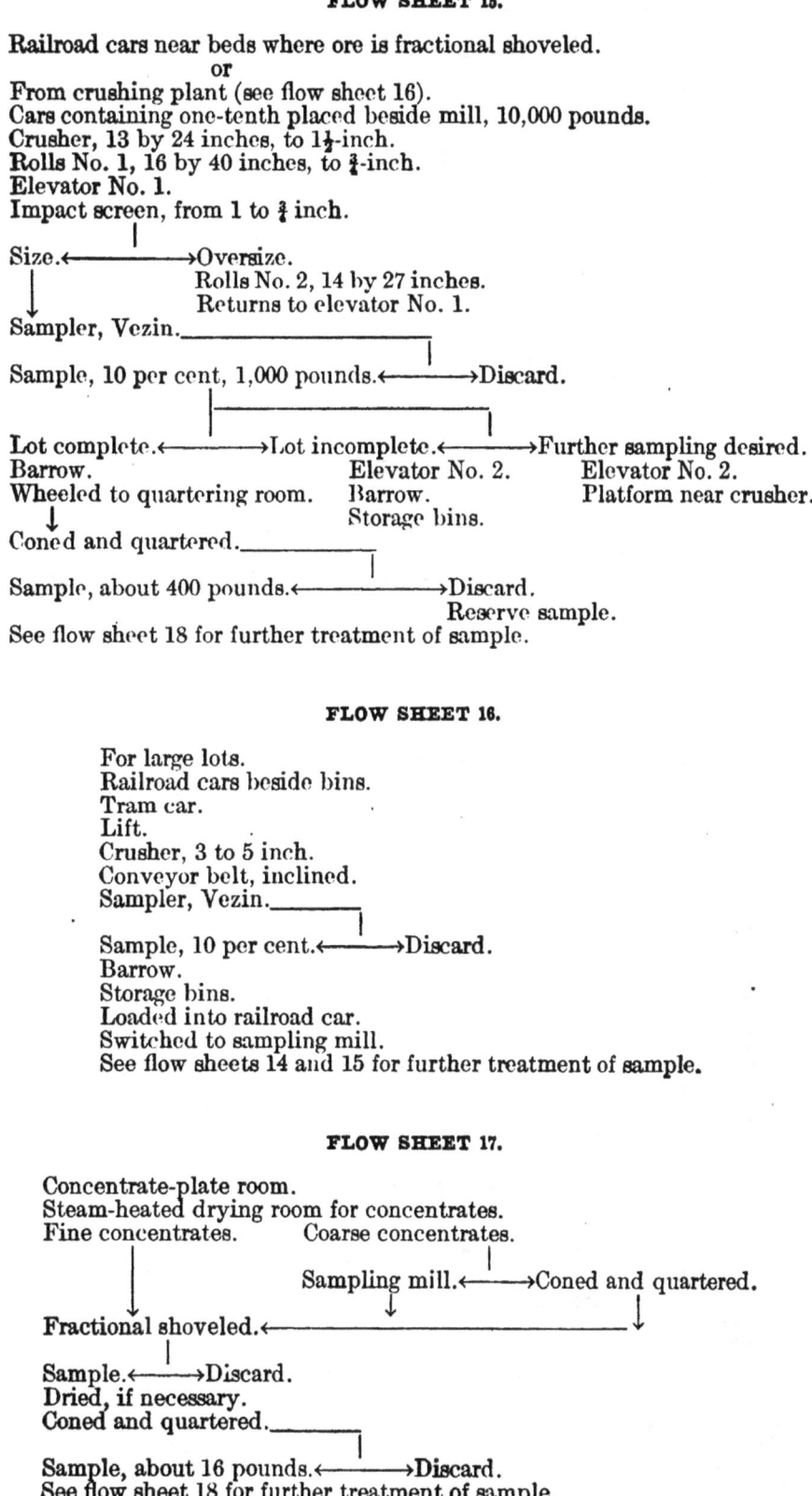

Size.◄————————►Oversize.
                Rolls No. 2, 14 by 27 inches.
                Returns to elevator No. 1.
Sampler, Vezin._____

Sample, 10 per cent, 1,000 pounds.◄———————►Discard.

Lot complete.◄————————►Lot incomplete.◄————————►Further sampling desired.
Barrow.   Elevator No. 2.   Elevator No. 2.
Wheeled to quartering room. Barrow.   Platform near crusher.
                        Storage bins.
Coned and quartered._____

Sample, about 400 pounds.◄————————►Discard.
                                    Reserve sample.
See flow sheet 18 for further treatment of sample.

FLOW SHEET 16.

For large lots.
Railroad cars beside bins.
Tram car.
Lift.
Crusher, 3 to 5 inch.
Conveyor belt, inclined.
Sampler, Vezin._____

Sample, 10 per cent.◄———————►Discard.
Barrow.
Storage bins.
Loaded into railroad car.
Switched to sampling mill.
See flow sheets 14 and 15 for further treatment of sample.

FLOW SHEET 17.

Concentrate-plate room.
Steam-heated drying room for concentrates.
Fine concentrates.   Coarse concentrates.

                        Sampling mill.◄———————►Coned and quartered.

Fractional shoveled.◄———————————————————————

Sample.◄———————►Discard.
Dried, if necessary.
Coned and quartered._____

Sample, about 16 pounds.◄———————►Discard.
See flow sheet 18 for further treatment of sample.

FLOW SHEET 18.

Sample from quartering floor.
All dried, if desired.
Hopper No. 1, flush with floor.
Electric crane hoist.
Rolls, 12 by 40 inches, to $\frac{1}{4}$-inch.
Barrow.
Returned to hopper No. 1 again.
Returned to electric crane hoist again.
Riffle; original and duplicate samples started here.

Sample.←————————————→Discard.
Returned to hopper No. 1 and reriffled until
Final sample is about 16 pounds.
Grinding room.
Steam drier.
Grinder, Englebach type, to about 60-mesh.
Revolving mixer, Anaconda type.
Riffle.

Sample, exactly 20 ounces.←————→Discard.
Hand sieve, 80 to 200 mesh.

Size.←————————————→Oversize.
Revolving mixer again.   Bucking board.
Coned and quartered on rubber sheet.   Returns to hand sieve.
All pulp put in four sacks.

FLOW SHEET 19.

Railroad cars at side of mill, 100,000 pounds.
Ore contains in gold

Over $2\frac{1}{2}$ ounces.←————→$2\frac{1}{2}$ ounces or less.
                Fractional shoveled.

                Sample, 20 per cent, 20,000 pounds.←————→Discard.
Crusher, to 1-inch.←—Center of car.   Beds.
Rolls No. 1, to $\frac{1}{4}$-inch.
Elevator No. 1.
Revolving screen No. 1, $\frac{1}{4}$-inch mesh.

Size.←————————————→Oversize.
Hopper No. 1.   Returns to elevator No. 1.
Hopper No. 2, entire lot held here.
Mill cleaned.
Elevator No. 1.
Revolving screen No. 1.
Hopper No. 1.
Sampler No. 1, Vezin.

Sample, 20 per cent, 4,000 pounds.←————→Discard.
Hopper No. 3, entire sample held here.
Mill cleaned.
Passed again through sampler No. 1, if desired.
Elevator No. 1.
Hopper No. 1, entire sample held here.
Barrows.
Lift No. 1.
Compound riffle; original and duplicate samples started here.

Sample, $12\frac{1}{2}$ per cent, 800 pounds.←————→Discard.
Barrows.
Lift No. 1.
Wheelbarrows.
To another part of mill.
Rolls No. 2, to $\frac{1}{8}$-inch.
Wheelbarrows.
Riffling room.
See flow sheet 20 for further treatment of sample.

FLOW SHEET 20.

Riffle, ore about ½-inch mesh.

Sample, about 20 pounds.←——→Discard.
Steam drier.
Grinder, Englebach type.
Riffle._____

Sample, 2 to 4 pounds.←——→Discard.
Hand sieve, 120 or 150 mesh.

Size.←————————————→Oversize.
Mixed on glass plate with spatula.   Bucking board.
                                      Returns to hand sieve.
Flattened into thin cake.
Part put in four sacks with spatula, balance discarded.

FLOW SHEET 21.

Cars on trestle.
Dumped or shoveled into bins.
Crusher No. 1, 10 by 20 inches, to 2-inch.
Conveyor belt.
Elevator No. 1._____

All to be sampled.←——→Coarse only to be sampled.
                       Revolving screen, ½ or ⅜-inch mesh.

Inclined chute.←————Oversize.←————→Size.
                                    Belt conveyor.
                                    Storage bins.

Sampler No. 1, Snyder._____

Sample, 25 per cent, 25,000 pounds.←——→Discard.
Elevator No. 2.
Crusher No. 2, to 1½-inch.
Sampler No. 2, Snyder._____

Sample, 20 per cent, 5,000 pounds.←——→Discard.
Rolls No. 1, to ¾-inch.
Sampler No. 3, Snyder._____

Sample, 16⅔ per cent, 833 pounds.←——→Discard.
Rolls No. 2, 10 by 24 inches, to ¼-inch.
Sampler No. 4, Snyder._____

Sample, 16⅔ per cent, 139 pounds.←——→Discard.
Through canvas spout.
Bucket.
Bucking room.
See flow sheet 23 for further treatment of sample.

FLOW SHEET 22.

High grades and concentrates, 50,000 pounds.
Unsacked in railroad car.
Fractional shoveled._____

Sample, 20 per cent.←————→Discard.
                           Bins.
Quartering room.
Coned and quartered._____

Sample, about 200 pounds.←——→Discard.
See flow sheet 23 for further treatment of sample. The whole lot is resampled the
  following day; on each sample there are 10 assay buttons made.

FLOW SHEET 23.

Riffle No. 1._____
                           |
Sample, about 10 pounds.←——→Discard.
         ↓   Held as reserve.
Steam drier.
Grinder, Englebach type, about 40-mesh.
Riffle No. 2._____
                           |
Sample, about 25 ounces.←——→Discard.
Hand sieve, 100, 120, and 150 mesh.
              |
Size.←——→Oversize.
    | Grinder, disk type.
    | Returns to hand sieve.
    ↓
Rolling cloth.
All pulp riffled to four parts.

FLOW SHEET 24.

Railroad cars over hopper.
Unloaded into hopper.
Shaking tray.
Crusher, 10 by 20 inches, to $1\frac{1}{2}$-inch.
Elevator No. 1.
Sampler No. 1, Brunton oscillating.___
                                      |
Sample, 20 per cent, 20,000 pounds.←——→Discard.
Shaking tray.
Rolls No. 1, 15 by 36 inches, to 1-inch.
Shaking tray.
Elevator No. 2.
Sampler No. 2, Brunton oscillating.___
                                      |
Sample, 20 per cent, 4,000 pounds.←——→Discard.
Shaking tray.
Rolls No. 2, 15 by 36 inches, to $\frac{1}{2}$-inch.
Sampler No. 3, Brunton oscillating.__
                                     |
Sample, 20 per cent, 800 pounds.←——→Discard.
Shaking tray.
Rolls No. 3, 10 by 20 inches, to $\frac{1}{4}$-inch.
Sampler No. 4, Brunton oscillating.__
                                     |
Sample, 20 per cent, 160 pounds.←——→Discard.
Shaking tray.
Rolls No. 4, 8 by 12 inches, to $\frac{1}{8}$-inch.
Sample bin.
See flow sheet 25 for further treatment of sample.

FLOW SHEET 25.

Riffle, Brunton type._____

Sample, from 25 to 40 pounds.←————————→Discard.
Moisture sample taken here.

Put in covered cans.
Taken to bucking room.
Steam drier.
Grinder, Englebach type, $\frac{1}{40}$-mesh.
Riffle, Brunton type._____

Sample, 3 to 4 pounds.←——————————→Discard.
Hand sieve, 100, 120, and 150 mesh.

Size._____Oversize.
Rolling cloth.   Bucking board.
All pulp riffled into four to ten sacks. Returns to hand sieve.

FLOW SHEET 26.

Cars on trestle over bins.
Unloaded into bins.
Conveyor belt.
Shaking grizzly.
Crusher No. 1, 12 by 24 inches, to $2\frac{1}{2}$-inch.
Conveyor belt.
Elevator No. 1.
Sampler No. 1, Brunton oscillating._____

Sample, 20 per cent, 20,000 pounds.←———→Discard.
Shaking tray.
Crusher No. 2, 10 by 20 inches, to $1\frac{1}{2}$-inch.
Sampler No. 2, Brunton oscillating.____

Sample, 20 per cent, 4,000 pounds.←————→Discard.
Shaking tray.
Rolls No. 1, 12 by 48 inches, to $\frac{3}{4}$-inch.
Sampler No. 3, Brunton oscillating.___

Sample, 20 per cent, 800 pounds.←————→Discard.
Shaking tray.
Rolls No. 2, 15 by 26 inches, to $\frac{1}{4}$-inch.
Sampler No. 4, Brunton oscillating.___

Sample, 20 per cent, 160 pounds.←————→Discard.
Sample bin.
Trammed to cutting room.
See flow sheet 27 for further treatment of sample.

FLOW SHEET 27.

From sample bin.
Iron plate.
Brunton one-half or quarter shovel.

Sample, about 150 pounds.←——————→Discard.
Riffle No. 1.

Sample, about 40 pounds.←——————→Discard.
Electric drier.
Grinder, Englebach type, to 60-mesh.
Riffle No. 2.

Sample, about 50 ounces.←——————→Discard.
Grinder, disk type.
Hand sieve.

Size.←——————————————→Oversize.
Revolving mixer, Anaconda type.   Bucking board.
All pulp riffled into four sacks.   Return to hand sieve.

FLOW SHEET 28.

Cars on trestle over bins.
Unloaded into bins.
Distributing car.
Bin of 50-ton capacity, 100,000 pounds.
Shaking tray.
Crusher No. 1, 12 by 24 inches, to 3-inch.
Conveyor belt.
Elevator No. 1.
Sampler No. 1, Brunton oscillating.

Sample, 20 per cent, 20,000 pounds.←——————→Discard.
Crusher No. 2, 5 by 15 inches (8 by 20 inch crusher in second unit).
Shaking tray.
Rolls No. 1, 15 by 40 inches, to ¾-inch.
Sampler No. 2, Brunton oscillating.

Sample, 20 per cent, 4,000 pounds.←——————→Discard.
Hopper.
Shaking tray.
Rolls No. 2, 14 by 27 inches, to ¼-inch.
Sampler No. 3, Brunton oscillating.

Sample, 20 per cent, 800 pounds.←——————→Discard.
Sample car on overhead track.
Taken to cutting room.
See flow sheet 29 for further treatment of sample.

FLOW SHEET 29.

Sample dumped on steel floor.
Brunton quarter shovel._____

|

Sample, about 30 pounds.←——→Discard.
Steam drier.
Grinder No. 1, Englebach type, to 20-mesh.
Riffle No. 1._____

|

Sample, about 3 pounds.←——→Discard.
Grinder No. 2, disk type, to 40-mesh.
Hand sieve, 100 to 120 mesh.←————————

|

Size.←————————————→Oversize.
Revolving mixer, Anaconda type. Bucking board.—
All pulp riffled into four sacks.

FLOW SHEET 30.

Railroad cars over bin.
Unloaded into bin.
Fed direct to grizzly
Crusher, 10 by 20 inches, to 2½-inch.
Belt conveyor.
Elevator No. 1.
Inclined spout.
Small hopper.
Sampler No. 1, Brunton oscillating._____

|

Sample, 20 per cent, 20,000 pounds.←——→Discard.
Small hopper.
Short shaking tray.
Rolls No. 1, 16 by 36 inches, to 1-inch.
Sampler No. 2, Brunton oscillating._____

|

Sample, 20 per cent, 4,000 pounds.←——→Discard.
Small hopper.
Short shaking tray.
Rolls No. 2, 14 by 30 inches, to ½-inch.
Sampler No. 3, Brunton oscillating._____

|

Sample, 20 per cent, 800 pounds.←——→Discard.
Small hopper.   Reserved, according to mining law
Short shaking tray.   of Montana.
Rolls No. 3, 12 by 20 inches, to ¼-inch.
Inclosed room.
Sampler No. 4, Brunton oscillating._____

|

Sample, 20 per cent, 160 pounds.←——→Discard.
Bucket on overhead track.
Locked cutting room.
See flow sheet 31 for further treatment of sample.

34915°—16——6

FLOW SHEET 31.

Sample dumped on steel floor.
Coned.
Brunton one-half or quarter shovel.

Sample, about 8 pounds.⟵————————————————⟶Discard.
Electric drier.
Grinder No. 1, Englebach type, to 60-mesh.
Riffle No. 1.

Sample, about 32 ounces.⟵————————————⟶Discard.
Grinder No. 2, disk type.
Hand sieve, 100 to 200 mesh.

Size.⟵————————————⟶Oversize.
Rolling cloth.   First residue to grinder No. 2, final residue to
                                                 bucking board.
All pulp riffled into four sacks.   Returned to hand sieve.

FLOW SHEET 32.

Railroad cars beside uncovered bedding bins.
Fractional shoveling in cars.

Sample, 10 per cent, 20,000 pounds.⟵————————⟶Discard.
Fractional shoveled in cars.

Sample, 33⅓ or 50 per cent, 6,667 or 10,000 pounds.⟵——⟶Discard.
Wheeled to plate room or to other mills.
See flow sheets 33, 34, 35, 36, 37.

FLOW SHEET 33.

Generally after fractional shoveling.
Cars switched to side of mill, 10,000 pounds.
Shoveled to grizzly.
Crusher No. 1, 9 by 15 inches, to 1½-inch.
Elevator No. 1.
Sampler No. 1, Brunton vibrating.

Sample, 20 per cent, 2,000 pounds.⟵——⟶Discard.
Crusher No. 2, 4 by 10 inches, to ¾-inch.
Sampler No. 2, Brunton vibrating.

Sample, 20 per cent, 400 pounds.⟵——⟶Discard.
Rolls, 12 by 20 inches, to ¾-inch.
Wheeled to plate room.
See flow sheet 36.

FLOW SHEET 34.

Generally after fractional shoveling.
Cars switched to side of mill, 10,000 pounds.
Wheelbarrows.
Crusher, 9 by 15 inches, to 1½-inch.
Elevator No. 1.
Rolls No. 1, 16 by 36 inches, to ¼-inch.
Elevator No. 2.
Revolving screen, ¼-inch holes.

Size.⟵————⟶Oversize.
  ↓   Rolls No. 2, 14 by 27 inches.
Hopper.   Returns to elevator No. 2.
Sampler No. 1, Brunton vibrating.

Sample, 20 per cent, 2,000 pounds.⟵——⟶Discard.
Sampler No. 2, Brunton vibrating.

Sample, 20 per cent, 400 pounds.⟵——⟶Discard.
Sample hopper in adjoining room.
Wheeled to plate room.

FLOW SHEET 35.

For large lots of three or more carloads, 150,000 pounds.
Railroad cars beside mill.
Shoveled to pan conveyor.
Grizzly of steel rails.
Crusher, gyratory, No. 5, to 1½-inch.
Inclined conveyor belt.
Hopper.____

Oxide ores.←——→Sulphide ores.
            Revolving screen, 2-mesh, about ½-inch.

←————Size.←——→Oversize.
            Revolving screen, two sections, 1-inch and 2-inch.
            ⅝ by 1 inch slots.

            Size.←————————————→Oversize.
            Rolls, 14 by 30 inches.   2-inch section.
            Returns to ½-inch revolv-
               ing screen.   Size.←———→Oversize.
                                      Rolls, 16 by   Crusher.
                                        36 inches.   Gyratory No. 4.
                                      Returns to   Returns to ½-inch
                                        ½-inch re-   revolving screen.
                                        vol ving
                                        screen.

Hopper.
Sampler No. 1, Vezin.____

Sample, 20 per cent, 30,000 pounds.←——→Discard.
Shaking tray.____

Sampler No. 2, special Vezin.←————→Sampler No. 5, Vezin.

Sample, 20 per cent, 6,000 pounds.←——→Discard.←——→Sample, 12½ per cent.
Shaking tray.   Reserved per Montana mining
Rolls, 12 by 12 inches, to ¼-inch.   law.
Sampler No. 3, Vezin.____

Sample, 20 per cent, 1,200 pounds.←——→Discard.
Shaking tray.
Sampler No. 4, Vezin.____

Sample, 20 per cent, 240 pounds.←——→Discard.
Locked sample hopper.
Wheeled in barrow to bucking room.
Dried, if necessary.
Coned and quartered.____

Sample, about 150 pounds.←————→Discard.
                               50 pounds held for resample.
See flow sheet 37 for further treatment of sample.

FLOW SHEET 36.

Plate room.
Ore dumped in ring on steel-plate floor.
Coned over wooden cross.
Drawn into cake and quarters removed with cross in place,
Process repeated until sample is reduced to about 150 pounds.

Sample, about 150 pounds.←——→Discard.
Wheeled to roll room.   Reserve sample taken in accordance with State
                                      mining law.
See flow sheet 37 for further treatment of sample.

FLOW SHEET 37.

Bucking room, about 150 pounds.

Riffle,

Sample, about 10 pounds.← →Discard.

Dried.
Grinder, Englebach type, to 50-mesh, check sample grabbed here.
Riffle.

Sample, about 2½ pounds.← →Discard.
Hand sieve, 100 or 120 mesh.

Size.← →Oversize.
Rolled on glazed paper.   Bucking board.
Put in six sacks with spatula, excess discarded.   Returned to hand sieve.

Roll room, about 150 pounds.
Dried, if necessary.
Rolls, 12 by 20 inches, to 20-mesh.
Riffle.

50 pounds reserved for resample.

FLOW SHEET 38.

For large lots, 500,000 pounds or more.
Grizzly.
Crusher No. 1, 24 by 36 inches, to 5½-inch.
Inclined conveyor belt.
Sampler No. 1, Vezin.

Sample, 10 per cent, 50,000 pounds.← →Discard.
Shaking tray.
Crusher No. 2, 15 by 24 inches, to 2-inch.
Angled spout.
Sampler No. 2, Vezin.

Sample, 20 per cent, 10,000 pounds.← →Discard.
Inclined conveyor belt.
Covered steel hopper.
Long shaking tray.
Rolls No. 1, 16 by 42 inches, between 1 and 1¾ inches.
Chute.
Inclined conveyor belt.
Hopper.
Sampler No. 3, Vezin.

Sample, 10 per cent, 1,000 pounds.← →Discard.
Long chute.
Conveyor belt.
Hopper.
Shaking tray.
Rolls No. 2, 12 by 20 inches, between ½ and ¾-inch.
Chute.
Inclined conveyor belt.
Hopper.
Sampler No. 4, Vezin.

Sample, 20 per cent, 200 pounds.← →Discard.
Long chute.
Conveyor belt.
Rolls No. 3, 12 by 20 inches, set close.
Hopper.
Compound riffles, on track.   Omitted as desired.

Sample, about 125 pounds.← →Discard.
Put in canvas sacks.
Taken to bucking room.
See flow sheet 39 for further treatment of sample.

FLOW SHEET 39.

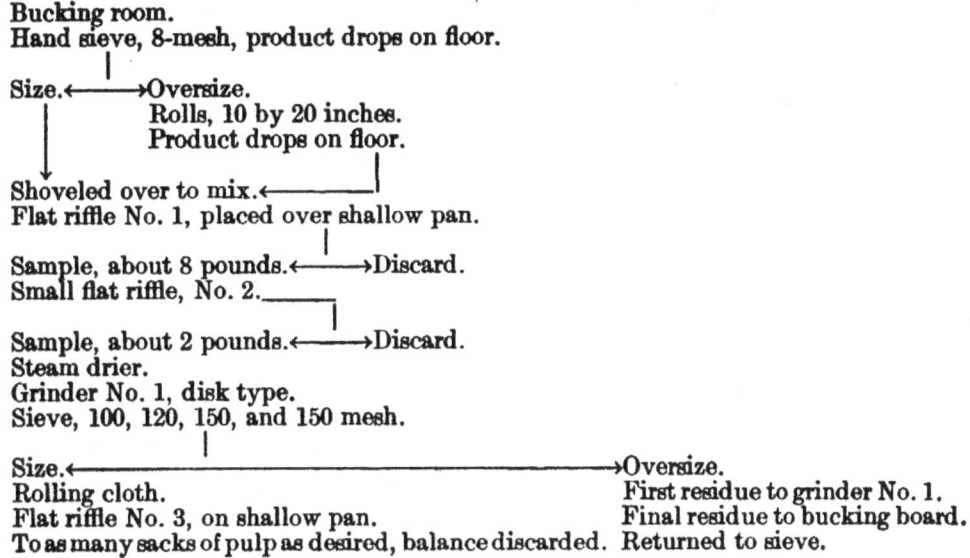

Bucking room.
Hand sieve, 8-mesh, product drops on floor.

Size.←———→Oversize.
　　　　　　　Rolls, 10 by 20 inches.
　　　　　　　Product drops on floor.

Shoveled over to mix.←——————
Flat riffle No. 1, placed over shallow pan.

Sample, about 8 pounds.←———→Discard.
Small flat riffle, No. 2._____

Sample, about 2 pounds.←———→Discard.
Steam drier.
Grinder No. 1, disk type.
Sieve, 100, 120, 150, and 150 mesh.

Size.←——————————————————→Oversize.
Rolling cloth.　　　　　　　　　　　First residue to grinder No. 1.
Flat riffle No. 3, on shallow pan.　　Final residue to bucking board.
To as many sacks of pulp as desired, balance discarded. Returned to sieve.

FLOW SHEET 40.

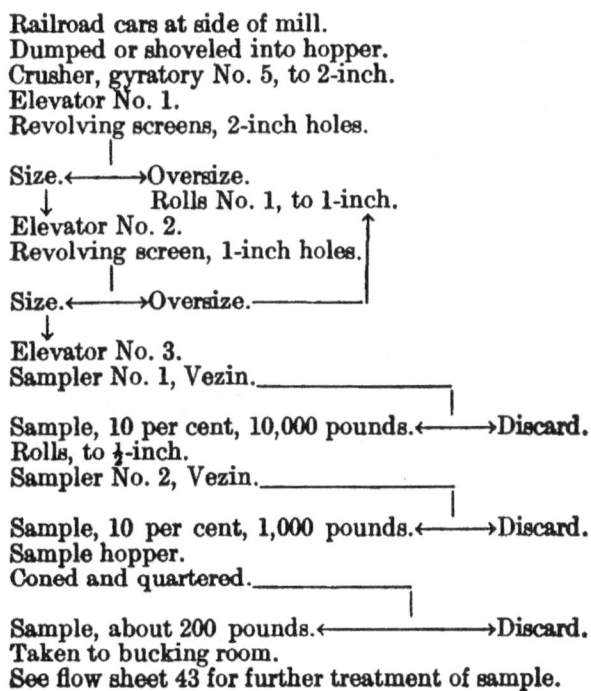

Railroad cars at side of mill.
Dumped or shoveled into hopper.
Crusher, gyratory No. 5, to 2-inch.
Elevator No. 1.
Revolving screens, 2-inch holes.

Size.←———→Oversize.
　　　↓　　　　Rolls No. 1, to 1-inch.
Elevator No. 2.
Revolving screen, 1-inch holes.

Size.←———→Oversize.————
　　　↓
Elevator No. 3.
Sampler No. 1, Vezin._____

Sample, 10 per cent, 10,000 pounds.←———→Discard.
Rolls, to ½-inch.
Sampler No. 2, Vezin._____

Sample, 10 per cent, 1,000 pounds.←———→Discard.
Sample hopper.
Coned and quartered._____

Sample, about 200 pounds.←————→Discard.
Taken to bucking room.
See flow sheet 43 for further treatment of sample.

FLOW SHEET 41.

Railroad cars on elevated track.
Unloaded into hopper.
Crusher No. 1, gyratory, to 2-inch.
Elevator No. 1.
Sampler No. 1, Vezin._____

Sample, 20 per cent, 20,000 pounds.←——→Discard.
Crusher No. 2, to 1-inch.
Sampler No. 2, Vezin._____

Sample, 20 per cent, 4,000 pounds.←——→Discard.
Rolls No. 1, to ½-inch.
Elevator No. 2.
Sampler No. 3, Vezin._____

Sample, 20 per cent, 800 pounds.←——→Discard.
Rolls No. 2, to ¼-inch.
Sampler No. 4, Vezin._____

Sample, 20 per cent, 160 pounds.←——→Discard.
To bucking room.
See flow sheet 43 for further treatment of sample.

FLOW SHEET 42.

Railroad cars at side of mill.
Wheelbarrows.
Crusher No. 1, 10 by 18 inches, to 2-inch.
Conveyor belt.
Sampler No. 1, Vezin._____

Sample, 20 per cent, 20,000 pounds.←——→Discard.
Shaking tray.
Crusher No. 2, to 1-inch.
Conveyor belt.
Sampler No. 2, Vezin._____

Sample, 20 per cent, 4,000 pounds.←——→Discard.
Shaking tray.
Rolls No. 1, to ½-inch.
Sampler No. 3, Vezin._____

Sample, 20 per cent, 800 pounds.←——→Discard.
Shaking tray.
Rolls No. 2, to ¼-inch.
Sampler No. 4, Vezin._____

Sample, 20 per cent, 160 pounds.←——→Discard.
Small sample box.
Taken to bucking room.
See flow sheet 43 for further treatment of sample.

FLOW SHEET 43.

Rolls, 12 by 20 inches, to $\frac{1}{4}$-inch, about 100 pounds.
Riffle No. 1, Brunton type.

Sample, about 12 pounds.————————→Discard.
Steam drier, short time.
Grinder, Englebach type, to 16-mesh.
Riffle No. 2, Brunton type.

Sample, about 4 pounds.————————→Discard.
Steam drier.
Grinder, Englebach type, to 40-mesh.
Riffle, Brunton type.

Sample, about 15 ounces.————————→Discard.
Hand sieve, 100, 120, or 150 mesh.   Reserve.

Size.←————————————→Oversize.
Mixed on whirling gold pans.   Bucking board.
All pulp sacked with spatula.   Returned to hand sieve.

FLOW SHEET 44.

Principally for large lots, 200,000 pounds.
Railroad cars over bins.
Crusher, gyratory type, to 3-inch.
Belt conveyor.
Sampler No. 1, Vezin.

Sample, 10 per cent, 20,000 pounds.←————→Discard.
Storage bins, or a chute.
Crusher No. 2, 14 by 24 inches, to 2½ or 2 inch.
Sampler No. 2, Vezin.

Sample, 20 per cent, 4,000 pounds.←————→Discard.
Shaking tray.
Rolls No. 1, 16 by 36 inches, to about 1½-inch.
Elevator No. 1.
Sampler No. 3, Vezin.

Sample, 10 per cent, 400 pounds.←————→Discard.
Rolls No. 2, 12 by 24 inches, to $\frac{3}{4}$ or $\frac{1}{2}$ inch.
Hoppers in adjoining room.
Recrushed in necessary.
Coned and quartered.

Sample, about 100 pounds.←————————→Discard.
To bucking room.
See flow sheet 45 for further treatment of sample.

FLOW SHEET 45.

Rolls, 12 by 20 inches, to $\frac{1}{8}$-inch, about 100 pounds.
Riffle No. 1, Brunton type.

Sample, about 12 pounds.⟵————————————⟶Discard.
Steam drier.
Grinder, Englebach type, to 16-mesh.
Riffle No. 2, Brunton type.

Sample, about 4 pounds.⟵————————————⟶Discard.
Steam drier.
Grinder, Englebach type, to 40-mesh.
Riffle, Brunton type.

Sample, about 15 ounces.⟵————————————⟶Discard.
Hand sieve, 100 to 150 mesh.   Reserve.

Size.⟵————————————⟶Oversize.
Mixed on whirling gold pans.   Bucking board.
All pulp sacked with a spatula.   Returned to hand sieve.

FLOW SHEET 46.

Fractional shoveled in railroad cars.

Sample, 10, 20, 33$\frac{1}{3}$, or 50 per cent.⟵————⟶Discard.
Left in railroad car.
Car switched to door of coning room.
Loaded into tram cars.

If sufficiently fine.⟵————————————⟶If coarse.
Trammed to coning floor.   Trammed to lift.
Shoveled into cone or long pile.   See flow sheet 47 for further treatment.
Fractional shoveled.

Sample, 50 per cent.⟵————————————⟶Discard.
Coned.
Fractional shoveled.

Sample, 50 per cent.⟵————————————⟶Discard.
Coned on wooden cross.
Quartered with cross in position.

Sample, 50 per cent.⟵————————————⟶Discard.
Fractional shoveling, and coning and quartering, are repeated until the desired weight
    is obtained. This depends on grade of ore. During the last quartering the four
    quarters are treated as four separate samples.
Taken to drier.
See flow sheet 49 for further treatment of sample.

FLOW SHEET 47.

Principally for ores from custom sampling works.
Railroad cars beside mill.
Fractional shoveled in car._____

Sample, 10 or 20 per cent.←——————————→Discard.
Trammed to lift.   Trammed to bins.
Elevated.
Dumped on iron plate No. 1.
Shoveled into crusher No. 1, 6 by 16 inches, to 1-inch.
Inclined iron plate.
Forms semicone on floor.
Fractional shoveled._____

Sample, 33⅓ to 50 per cent.←——————————→Discard.
Taken to lift.   Trammed to bins.
Elevated.
Dumped on iron plate No. 2.
Shoveled into crusher No. 2, to ½-inch.
Falls into pile on floor.
Fractional shoveled._____

Sample, 50 per cent.←——————————→Discard.
Taken to lift.   Trammed to bins.
Rolls No. 1, 8 by 18 inches, to ¼-inch.
Falls into tramcar.
Taken to adjoining room.
Placed in can.
Rolls No. 2, 8 by 12 inches, to 8-mesh.
Into small car.
Dumped on iron plate.
Coned to mix.
Riffle No. 1, Brunton type._____

Sample, 50 per cent.←——————————→Sample, 50 per cent.
Riffle No. 1, Brunton type.   Riffle No. 1, Brunton type.

Sample A, 25⎱ ⎰Sample B, 25   Sample C, 25⎱ ⎰Sample D, 25
  per cent.⎰ ⎱ per cent.   per cent.⎰ ⎱ per cent.
A, B, C, and D, treated as separate samples.
Put on drier.
See flow sheet 49 for further treatment of sample.

FLOW SHEET 48.

For handling gold precipitates, high-grade slag refineries, and sweepings from assay
    offices, jeweler shops, and dental parlors, etc.
If the sample contains much refuse, it is burned before being weighed.
Ground in Chilean mill to pass through screens of varying mesh. Smelter and refinery
    products are ground to 10-mesh; custom material to 30-mesh; certain high-grade
    gold ores to 100-mesh.
Coned, ringed, coned and quartered.
Metallics weighed, melted into bar, and "shot" sampled.
Final sample treated as shown in flow sheet 49.

FLOW SHEET 49.

From drier.
To quartering floor.
Wooden-frame hand screen, 20 mesh.

Size.←————————————→Oversize.
                              Grinder, Englebach type.
                              Returned to 20-mesh screen.

Coned and quartered._____

Sample, to about 5 pounds.←————————→Discard.
Hand screen, 10-mesh.

Size.←————————————→Oversize.
                              Grinder, Englebach type.
                              Returned to 10-mesh screen.

Riffle, Brunton type._____

Sample, about 20 ounces.←————————→Discard.
Hand screen, 100 to 150 mesh.

Size.←————————→Oversize.
Rolled on manila paper.   Grinder, Englebach type.
Drier.   Returned to 100-mesh screen.
To another room.
Rerolled on manila paper.
Spread into thin cake.
Put into required number of sacks with spatula.
Balance of pulp sacked as reserve.

FLOW SHEET 50.

Railroad cars over ore hoppers, 100,000 pounds.
Shaking grizzly.
Crusher, 10 by 20 inches, to 2½-inch.
Shaking tray.
Elevator.
Sampler No. 1, Brunton oscillating.____

Sample, 20 per cent, 20,000 pounds.←————→Discard.
Shaking tray.
Roll No. 1, 16 by 36 inches, to 1-inch.
Sampler No. 2, Brunton oscillating.___

Sample, 20 per cent, 4,000 pounds.←————→Discard.
Shaking tray.
Rolls No. 2, 15 by 27 inches, to ¾-inch.
Sampler No. 3, Brunton oscillating.___

Sample, 20 per cent, 800 pounds.←————→Discard.
Shaking tray.
Rolls No. 3, 8 by 20 inches, to ½-inch.
Sampler No. 4, Brunton oscillating.___

Sample, 20 per cent, 160 pounds.←————→Discard.
Shaking tray.
Rolls No. 4, 8 by 12 inches, to 14-mesh.
Sample buggy in locked bin.
To cutting room.
See flow sheet 51 for further treatment of sample.

FLOW SHEET 51.

From sample buggy.
Riffle No. 1, Brunton type._____

Sample, 10 to 12 pounds.←——————————→Discard.
Electric drier.
Grinder, Englebach type, to 50-mesh.
Riffle No. 2, Brunton type._____

Sample, 20 to 24 ounces.←——————————→Discard.
Hand sieve, 120-mesh.

Size.←————————→Oversize.
Rolling cloth.   Bucking board.
                     Returned to 120-mesh sieve.
Riffle No. 3, Brunton type.
All pulp divided into four samples.
All pulp sacked in four sacks.

FLOW SHEET 52.

Ore received in storage bins.
Loaded by chutes into tramcars, 100,000 pounds.
Crusher, 8 by 12 inches, to 1½-inch.
Elevator No. 1.
Rolls No. 1, 16 by 36 inches, to 1-inch.
Elevator No. 2.
Revolving screen, 1-inch holes.

Size.←————————————→Oversize.
Bottomless hopper.   Returns to rolls No. 1.
Sampler, whistle-pipe type, 5 one-half sections.

Sample, one thirty-second, 3,125 pounds.←————→Discard.
Rolls No. 2, 8 by 12 inches, to ½-inch.
Wheelbarrow.
To quartering room.
Coned and quartered._____

Sample, 30 to 40 pounds.←————————————→Discard.
See flow sheet 53 for further treatment of sample.

FLOW SHEET 53.

From coning and quartering sample.
Riffle, flat type._____

Sample, 4 to 6 pounds.←——————————→Discard.
Steam drier.
Grinder, Englebach type, 30 to 40 mesh.
Hand sieve, 150-mesh.

Size.←————————→Oversize.
Rolled on paper.   Bucking board.
Sacked with spatula.   Returned to sieve.

FLOW SHEET 54.

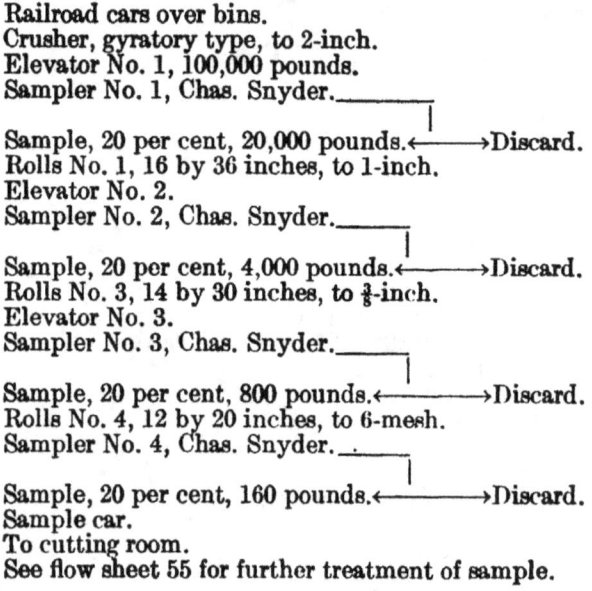

Railroad cars over bins.
Crusher, gyratory type, to 2-inch.
Elevator No. 1, 100,000 pounds.
Sampler No. 1, Chas. Snyder.

Sample, 20 per cent, 20,000 pounds.←——→Discard.
Rolls No. 1, 16 by 36 inches, to 1-inch.
Elevator No. 2.
Sampler No. 2, Chas. Snyder.

Sample, 20 per cent, 4,000 pounds.←——→Discard.
Rolls No. 3, 14 by 30 inches, to ¾-inch.
Elevator No. 3.
Sampler No. 3, Chas. Snyder.

Sample, 20 per cent, 800 pounds.←——→Discard.
Rolls No. 4, 12 by 20 inches, to 6-mesh.
Sampler No. 4, Chas. Snyder.

Sample, 20 per cent, 160 pounds.←——→Discard.
Sample car.
To cutting room.
See flow sheet 55 for further treatment of sample.

FLOW SHEET 55.

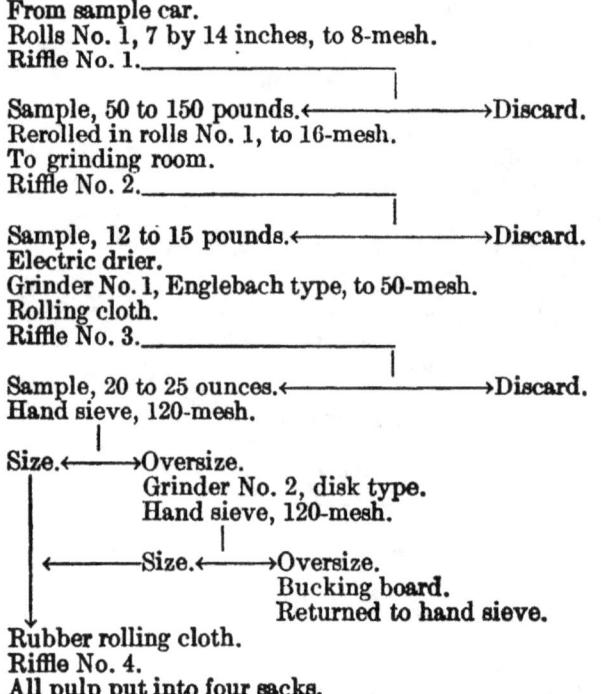

From sample car.
Rolls No. 1, 7 by 14 inches, to 8-mesh.
Riffle No. 1.

Sample, 50 to 150 pounds.←——→Discard.
Rerolled in rolls No. 1, to 16-mesh.
To grinding room.
Riffle No. 2.

Sample, 12 to 15 pounds.←——→Discard.
Electric drier.
Grinder No. 1, Englebach type, to 50-mesh.
Rolling cloth.
Riffle No. 3.

Sample, 20 to 25 ounces.←——→Discard.
Hand sieve, 120-mesh.

Size.←——→Oversize.
              Grinder No. 2, disk type.
              Hand sieve, 120-mesh.

    ←————Size.——→Oversize.
              Bucking board.
              Returned to hand sieve.
Rubber rolling cloth.
Riffle No. 4.
All pulp put into four sacks.

PUBLICATIONS ON MINE ACCIDENTS AND METHODS OF METAL MINING.

Limited editions of the following Bureau of Mines publications are temporarily available for free distribution. Requests for all publications can not be granted, and applicants should select only those publications that are of especial interest to them. All requests for publications should be addressed to the Director, Bureau of Mines, Washington, D. C.

BULLETIN 48. The selection of explosives used in engineering and mining operations, by Clarence Hall and S. P. Howell. 1913. 50 pp., 3 pls., 7 figs.

BULLETIN 53. Mining and treatment of feldspar and kaolin in the southern Appalachian region, by A. S. Watts. 1913. 170 pp., 16 pls., 12 figs.

BULLETIN 62. National mine rescue and first-aid conference, Pittsburgh, Pa., September 23–26, 1912, by H. M. Wilson. 1913. 74 pp.

BULLETIN 75. Rules and regulations for metal mines, by W. R. Ingalls, James Douglas, J. R. Finlay, J. Parke Channing, and John Hays Hammond. 1915. 296 pp., 1 fig.

BULLETIN 80. A primer on explosives for metal miners and quarrymen, by C. E. Munroe and Clarence Hall. 1915. 125 pp., 15 pls., 17 figs.

BULLETIN 101. Abstract of current decisions on mines and mining, October, 1914, to April, 1915, by J. W. Thompson. 1915. 138 pp.

BULLETIN 113. Abstracts of current decisions on mines and mining, reported from May to September, 1915, by J. W. Thompson. 1916. 124 pp.

TECHNICAL PAPER 4. The electrical section of the Bureau of Mines, its purpose and equipment, by H. H. Clark. 1911. 12 pp.

TECHNICAL PAPER 6. The rate of burning of fuse as influenced by temperature and pressure, by W. O. Snelling and W. C. Cope. 1912. 28 pp.

TECHNICAL PAPER 7. Investigations of fuse and miners' squibs, by Clarence Hall and S. P. Howell. 1912. 19 pp.

TECHNICAL PAPER 11. The use of mice and birds for detecting carbon monoxide after mine fires and explosions, by G. A. Burrell. 1912. 15 pp.

TECHNICAL PAPER 13. Gas analysis as an aid in fighting mine fires, by G. A. Burrell and F. M. Seibert. 1912. 16 pp., 1 fig.

TECHNICAL PAPER 15. An electrolytic method of preventing corrosion of iron and steel, by J. K. Clement and L. V. Walker. 1913. 19 pp., 10 figs.

TECHNICAL PAPER 17. The effect of stemming on the efficiency of explosives, by W. O. Snelling and Clarence Hall. 1912. 20 pp., 11 figs.

TECHNICAL PAPER 18. Magazines and thaw houses for explosives, by Clarence Hall and S. P. Howell. 1912. 34 pp., 1 pl., 5 figs.

TECHNICAL PAPER 19. The factor of safety in mine electrical installations, by H. H. Clark. 1912. 14 pp.

TECHNICAL PAPER 22. Electrical symbols for mine maps, by H. H. Clark. 1912. 11 pp., 8 figs.

TECHNICAL PAPER 9. Training with mine rescue breathing apparatus, by J. W. Paul. 1912. 16 pp.

TECHNICAL PAPER 30. Mine accident prevention at Lake Superior iron mines, by D. E. Woodbridge. 1913. 38 pp., 9 figs.

TECHNICAL PAPER 40. Metal-mine accidents in the United States during the calendar year 1911, compiled by A. H. Fay. 1913. 54 pp.

TECHNICAL PAPER 41. Mining and treatment of lead and zinc ores in the Joplin district, Missouri; a preliminary report, by C. A. Wright. 1913. 43 pp., 5 figs.

TECHNICAL PAPER 46. Quarry accidents in the United States during the calendar year 1911, compiled by A. H. Fay. 1913. 32 pp.

TECHNICAL PAPER 47. Portable electric mine lamps, by H. H. Clark. 1913. 13 pp.

TECHNICAL PAPER 58. The action of acid mine water on the insulation of electric conductors; a preliminary report, by H. H. Clark and L. C. Ilsley. 1913. 26 pp., 1 fig.

TECHNICAL PAPER 59. Fires in Lake Superior iron mines, by Edwin Higgins. 1913. 34 pp., 2 pls.

TECHNICAL PAPER 61. Metal-mine accidents in the United States during the calendar year 1912, compiled by A. H. Fay. 1913. 76 pp., 1 fig.

TECHNICAL PAPER 62. Relative effects of carbon monoxide on small animals, by G. A. Burrell, F. M. Seibert, and I. W. Robertson. 1914. 23 pp.

TECHNICAL PAPER 67. Mine signboards, by Edwin Higgins and Edward Steidle. 1913. 15 pp., 1 pl., 4 figs.

TECHNICAL PAPER 77. Report of the Committee on Resuscitation from Mine Gases, by W. B. Cannon, G. W. Crile, Joseph Erlanger, Yandell Henderson, and S. J. Meltzer. 1914. 36 pp., 4 figs.

TECHNICAL PAPER 92. Quarry accidents in the United States during the calendar year 1913, compiled by A. H. Fay. 1914. 76 pp.

TECHNICAL PAPER 94. Metal-mine accidents in the United States during the calendar year 1913, compiled by A. H. Fay. 1914. 73 pp.

TECHNICAL PAPER 105. Pulmonary disease among miners in the Joplin district, Missouri, and its relation to rock dust in the mines; a preliminary report, by A. J. Lanza and Edwin Higgins. 1915. 48 pp., 5 pls., 4 figs.

TECHNICAL PAPER 116. Miners' wash and change houses, by J. H. White. 1915. 27 pp., 3 pls., 3 figs.

TECHNICAL PAPER 129. Metal-mine accidents in the United States during the calendar year 1914, compiled by A. H. Fay. 1916. 96 pp., 3 pls., 1 fig.

MINERS' CIRCULAR 5. Electrical accidents in mines, their causes and prevention, by H. H. Clark, W. D. Roberts, L. C. Ilsley, and H. F. Randolph. 1911. 10 pp., 3 pls.

MINERS' CIRCULAR 8. First-aid instructions for miners, by M. W. Glasgow, W. A. Raudenbush, and C. O. Roberts. 1913. 67 pp., 51 figs.

MINERS' CIRCULAR 10. Mine fires and how to fight them, by J. W. Paul. 1912. 14 pp.

MINERS' CIRCULAR 13. Safety in tunneling, by D. W. Brunton and J. A. Davis. 1913. 19 pp.

MINERS' CIRCULAR 15. Rules for mine rescue and first-aid field contests, by J. W. Paul. 1913. 12 pp.

MINERS' CIRCULAR 17. Accidents from falls of rock or ore, by Edwin Higgins. 1914. 15 pp., 8 figs.

MINERS' CIRCULAR 20. How a miner can avoid some dangerous diseases, by A. J. Lanza and J. H. White. 1916. 26 pp., 4 figs.

INDEX.

96   INDEX.

66

A. CONE OF CRUSHED ORE WITH PARTICLES OF VARYING SIZE.

Note segregation of coarse from fine particles.

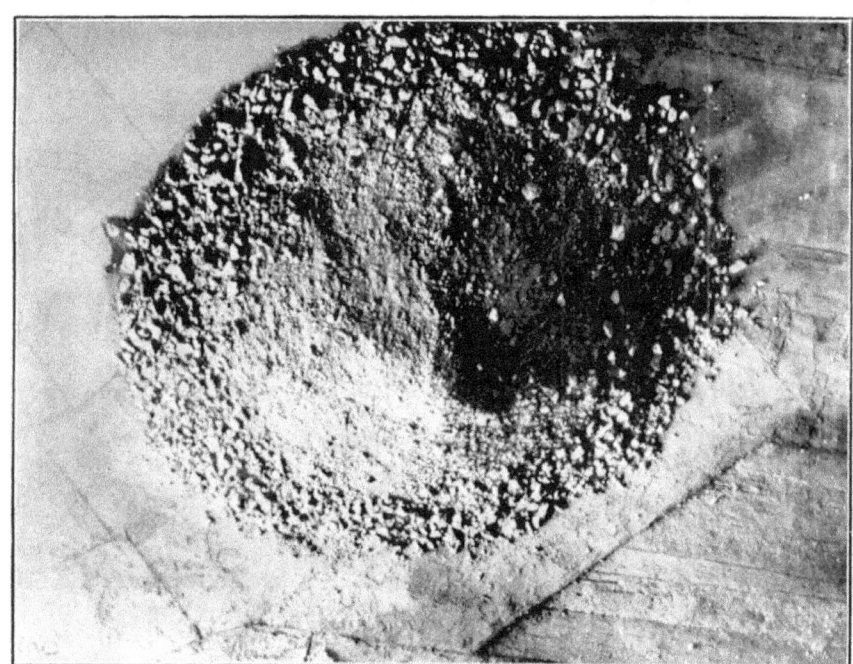

B. CONE OF CRUSHED ORE, AS VIEWED FROM ABOVE.

C. CONE AT BEGINNING OF FLATTENING PROCESS.

A. CONE OF CRUSHED ORE WITH PARTICLES MORE UNIFORM THAN THOSE OF CONE
SHOWN IN PLATE I, *A.*

B. PARTLY FLATTENED CONE.

C. CAKE FROM WHICH REJECT QUARTERS HAVE BEEN REMOVED

D. Riffle, Brunton type.   Side view.

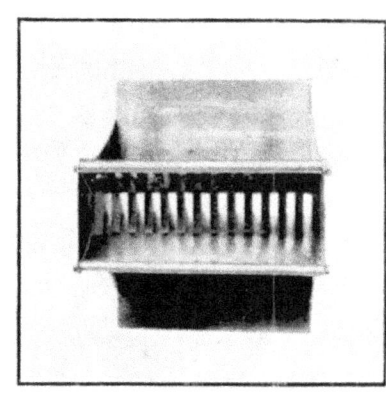

E. Riffle, Brunton type.   Top view.

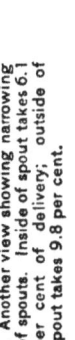

C. Another view showing narrowing of spouts.   Inside of spout takes 6.1 per cent of delivery;   outside of spout takes 9.8 per cent.

B. View showing narrowing of the spouts near the center.

A. Exterior of incorrectly constructed Vezin sampler.

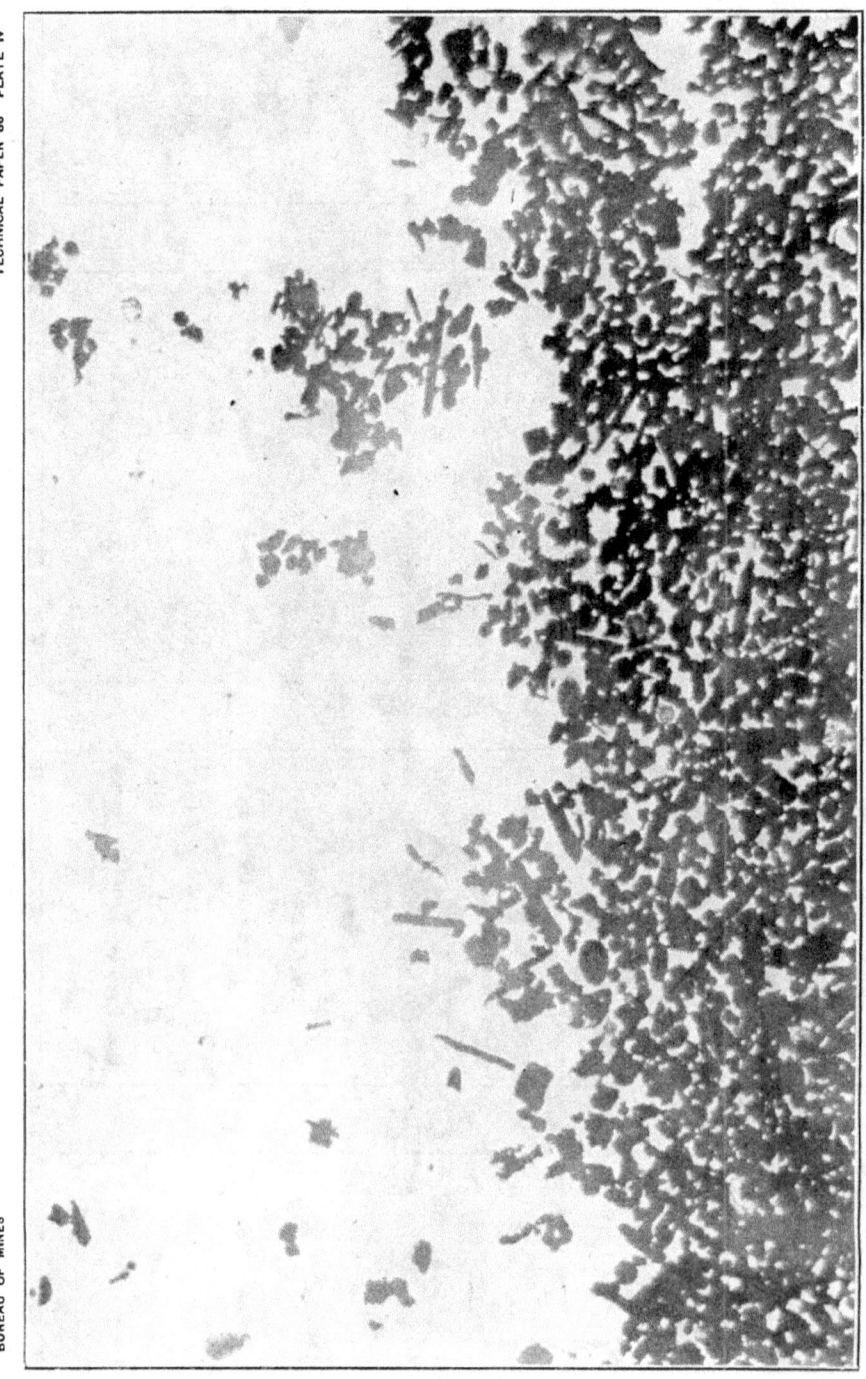

GOLD PARTICLES FROM PULP PASSED THROUGH 120-MESH SIEVE. GREATLY MAGNIFIED.

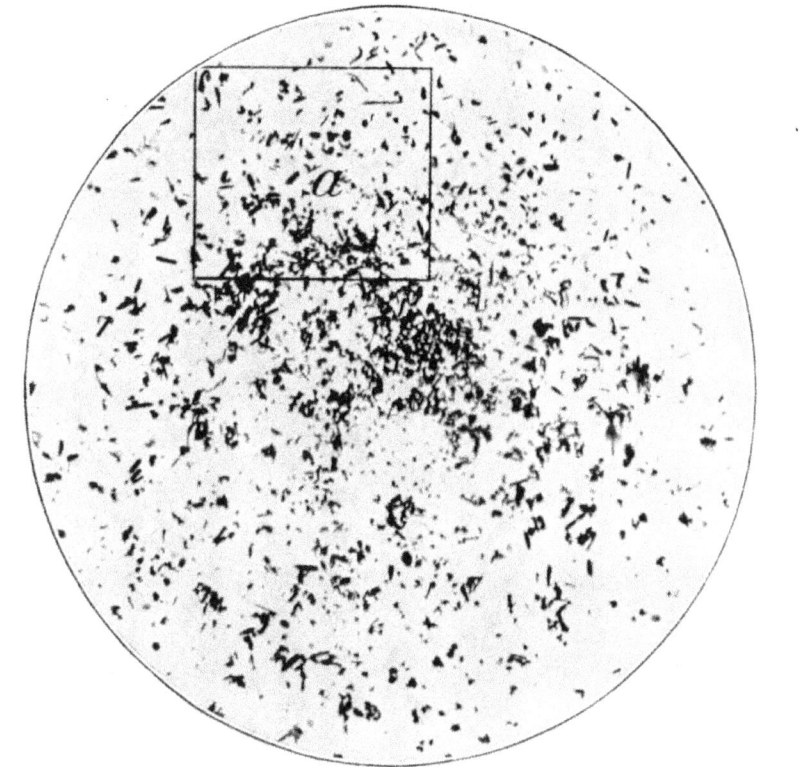

A. GOLD PARTICLES FROM PULP PASSED THROUGH 120-MESH SIEVE.   MAGNIFIED 6½ TIMES.

B. SECTION *a* OF PLATE V, *A.* REVERSED AND GREATLY MAGNIFIED.

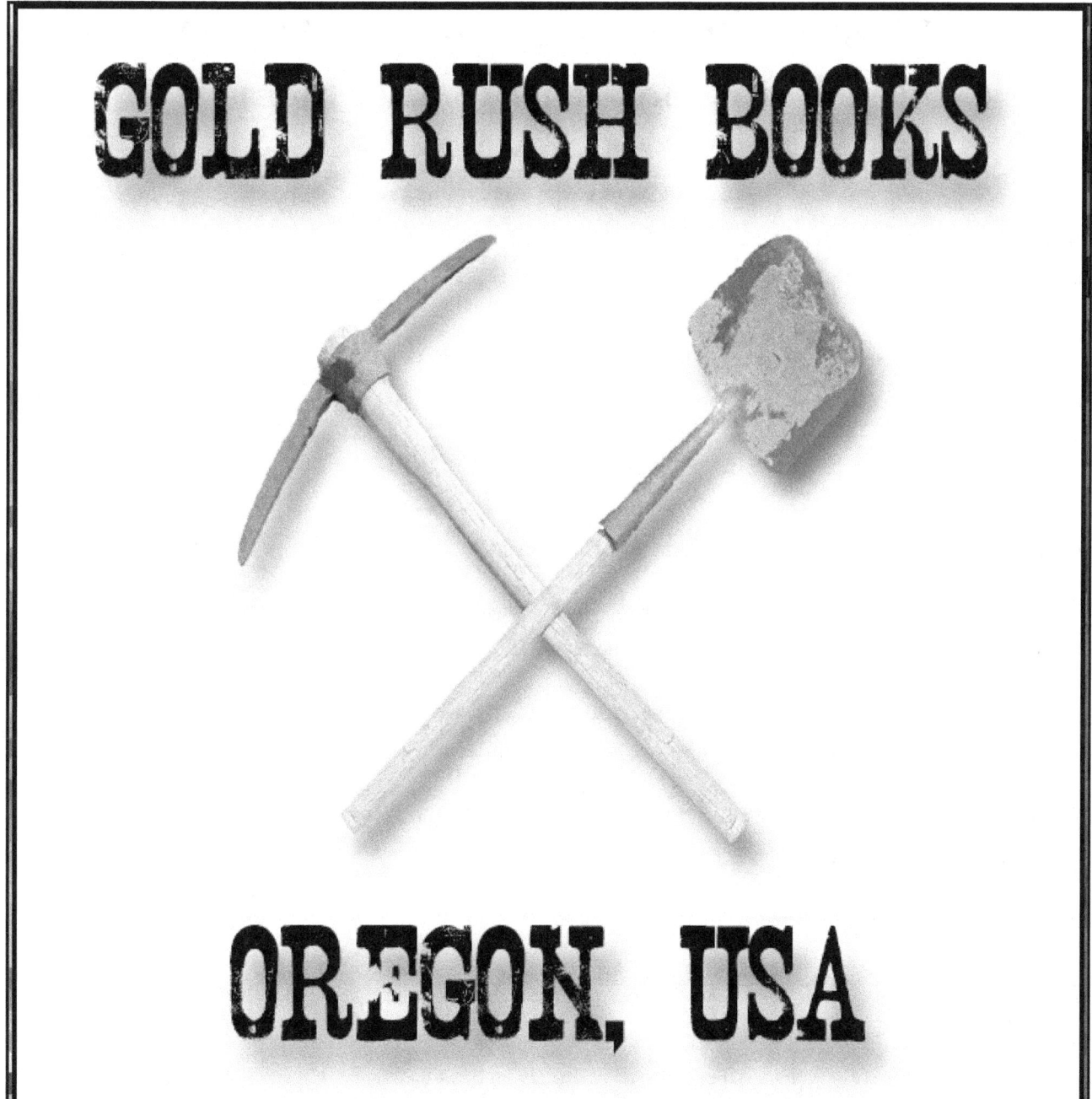

www.GoldMiningBooks.com

Books On Mining

Visit: www.goldminingbooks.com to order your copies or ask your favorite book seller to offer them.

Mining Books by Kerby Jackson

<u>Gold Dust: Stories From Oregon's Mining Years</u> - Oregon mining historian and prospector, Kerby Jackson, brings you a treasure trove of seventeen stories on Southern Oregon's rich history of gold prospecting, the prospectors and their discoveries, and the breathtaking areas they settled in and made homes. **5" X 8", 98 ppgs. Retail Price: $11.99**

<u>The Golden Trail: More Stories From Oregon's Mining Years</u> - In his follow-up to "Gold Dust: Stories of Oregon's Mining Years", this time around, Jackson brings us twelve tales from Oregon's Gold Rush, including the story about the first gold strike on Canyon Creek in Grant County, about the old timers who found gold by the pail full at the Victor Mine near Galice, how Iradel Bray discovered a rich ledge of gold on the Coquille River during the height of the Rogue River War, a tale of two elderly miners on the hunt for a lost mine in the Cascade Mountains, details about the discovery of the famous Armstrong Nugget and others. **5" X 8", 70 ppgs. Retail Price: $10.99**

Oregon Mining Books

<u>Geology and Mineral Resources of Josephine County, Oregon</u> - Unavailable since the 1970's, this important publication was originally compiled by the Oregon Department of Geology and Mineral Industries and includes important details on the economic geology and mineral resources of this important mining area in South Western Oregon. Included are notes on the history, geology and development of important mines, as well as insights into the mining of gold, copper, nickel, limestone, chromium and other minerals found in large quantities in Josephine County, Oregon. **8.5" X 11", 54 ppgs. Retail Price: $9.99**

<u>Mines and Prospects of the Mount Reuben Mining District</u> - Unavailable since 1947, this important publication was originally compiled by geologist Elton Youngberg of the Oregon Department of Geology and Mineral Industries and includes detailed descriptions, histories and the geology of the Mount Reuben Mining District in Josephine County, Oregon. Included are notes on the history, geology, development and assay statistics, as well as underground maps of all the major mines and prospects in the vicinity of this much neglected mining district. **8.5" X 11", 48 ppgs. Retail Price: $9.99**

<u>The Granite Mining District</u> - Notes on the history, geology and development of important mines in the well known Granite Mining District which is located in Grant County, Oregon. Some of the mines discussed include the Ajax, Blue Ribbon, Buffalo, Continental, Cougar-Independence, Magnolia, New York, Standard and the Tillicum. Also included are many rare maps pertaining to the mines in the area. **8.5" X 11", 48 ppgs. Retail Price: $9.99**

<u>Ore Deposits of the Takilma and Waldo Mining Districts of Josephine County, Oregon</u> - The Waldo and Takilma mining districts are most notable for the fact that the earliest large scale mining of placer gold and copper in Oregon took place in these two areas. Included are details about some of the earliest large gold mines in the state such as the Llano de Oro, High Gravel, Cameron, Platerica, Deep Gravel and others, as well as copper mines such as the famous Queen of Bronze mine, the Waldo, Lily and Cowboy mines. This volume also includes six maps and 20 original illustrations. **8.5" X 11", 74 ppgs. Retail Price: $9.99**

<u>Metal Mines of Douglas, Coos and Curry Counties, Oregon</u> - Oregon mining historian Kerby Jackson introduces us to a classic work on Oregon's mining history in this important re-issue of Bulletin 14C Volume 1, otherwise known as the Douglas, Coos & Curry Counties, Oregon Metal Mines Handbook. Unavailable since 1940, this important publication was originally compiled by the Oregon Department of Geology and Mineral Industries includes detailed descriptions, histories and the geology of over 250 metallic mineral mines and prospects in this rugged area of South West Oregon. **8.5" X 11", 158 ppgs. Retail Price: $19.99**

Metal Mines of Jackson County, Oregon - Unavailable since 1943, this important publication was originally compiled by the Oregon Department of Geology and Mineral Industries includes detailed descriptions, histories and the geology of over 450 metallic mineral mines and prospects in Jackson County, Oregon. Included are such famous gold mining areas as Gold Hill, Jacksonville, Sterling and the Upper Applegate. **8.5" X 11", 220 ppgs. Retail Price: $24.99**

Metal Mines of Josephine County, Oregon - Oregon mining historian Kerby Jackson introduces us to a classic work on Oregon's mining history in this important re-issue of Bulletin 14C, otherwise known as the Josephine County, Oregon Metal Mines Handbook. Unavailable since 1952, this important publication was originally compiled by the Oregon Department of Geology and Mineral Industries includes detailed descriptions, histories and the geology of over 500 metallic mineral mines and prospects in Josephine County, Oregon. **8.5" X 11", 250 ppgs. Retail Price: $24.99**

Metal Mines of North East Oregon - Oregon mining historian Kerby Jackson introduces us to a classic work on Oregon's mining history in this important re-issue of Bulletin 14A and 14B, otherwise known as the North East Oregon Metal Mines Handbook. Unavailable since 1941, this important publication was originally compiled by the Oregon Department of Geology and Mineral Industries and includes detailed descriptions, histories and the geology of over 750 metallic mineral mines and prospects in North Eastern Oregon. **8.5" X 11", 310 ppgs. Retail Price: $29.99**

Metal Mines of North West Oregon - Oregon mining historian Kerby Jackson introduces us to a classic work on Oregon's mining history in this important re-issue of Bulletin 14D, otherwise known as the North West Oregon Metal Mines Handbook. Unavailable since 1951, this important publication was originally compiled by the Oregon Department of Geology and Mineral Industries and includes detailed descriptions, histories and the geology of over 250 metallic mineral mines and prospects in North Western Oregon. **8.5" X 11", 182 ppgs. Retail Price: $19.99**

Mines and Prospects of Oregon - Mining historian Kerby Jackson introduces us to a classic mining work by the Oregon Bureau of Mines in this important re-issue of The Handbook of Mines and Prospects of Oregon. Unavailable since 1916, this publication includes important insights into hundreds of gold, silver, copper, coal, limestone and other mines that operated in the State of Oregon around the turn of the 19th Century. Included are not only geological details on early mines throughout Oregon, but also insights into their history, production, locations and in some cases, also included are rare maps of their underground workings. **8.5" X 11", 314 ppgs. Retail Price: $24.99**

Lode Gold of the Klamath Mountains of Northern California and South West Oregon
(See California Mining Books)

Mineral Resources of South West Oregon - Unavailable since 1914, this publication includes important insights into dozens of mines that once operated in South West Oregon, including the famous gold fields of Josephine and Jackson Counties, as well as the Coal Mines of Coos County. Included are not only geological details on early mines throughout South West Oregon, but also insights into their history, production and locations. **8.5" X 11", 154 ppgs. Retail Price: $11.99**

Chromite Mining in The Klamath Mountains of California and Oregon
(See California Mining Books)

Southern Oregon Mineral Wealth - Unavailable since 1904, this rare publication provides a unique snapshot into the mines that were operating in the area at the time. Included are not only geological details on early mines throughout South West Oregon, but also insights into their history, production and locations. Some of the mining areas include Grave Creek, Greenback, Wolf Creek, Jump Off Joe Creek, Granite Hill, Galice, Mount Reuben, Gold Hill, Galls Creek, Kane Creek, Sardine Creek, Birdseye Creek, Evans Creek, Foots Creek, Jacksonville, Ashland, the Applegate River, Waldo, Kerby and the Illinois River, Althouse and Sucker Creek, as well as insights into local copper mining and other topics. **8.5" X 11", 64 ppgs. Retail Price: $8.99**

Geology and Ore Deposits of the Takilma and Waldo Mining Districts - Unavailable since the 1933, this publication was originally compiled by the United States Geological Survey and includes details on gold and copper mining in the Takilma and Waldo Districts of Josephine County, Oregon. The Waldo and Takilma mining districts are most notable for the fact that the earliest large scale mining of placer gold and copper in Oregon took place in these two areas. Included in this report are details about some of the earliest large gold mines in the state such as the Llano de Oro, High Gravel, Cameron, Platerica, Deep Gravel and others, as well as copper mines such as the famous Queen of Bronze mine, the Waldo, Lily and Cowboy mines. In addition to geological examinations, insights are also provided into the production, day to day operations and early histories of these mines, as well as calculations of known mineral reserves in the area. This volume also includes six maps and 20 original illustrations. **8.5" X 11", 74 ppgs. Retail Price: $9.99**

Gold Mines of Oregon - Oregon mining historian Kerby Jackson introduces us to a classic work on Oregon's mining history in this important re-issue of Bulletin 61, otherwise known as "Gold and Silver In Oregon". Unavailable since 1968, this important publication was originally compiled by geologists Howard C. Brooks and Len Ramp of the Oregon Department of Geology and Mineral Industries and includes detailed descriptions, histories and the geology of over 450 gold mines Oregon. Included are notes on the history, geology and gold production statistics of all the major mining areas in Oregon including the Klamath Mountains, the Blue Mountains and the North Cascades. While gold is where you find it, as every miner knows, the path to success is to prospect for gold where it was previously found. **8.5" X 11", 344 ppgs. Retail Price: $24.99**

Mines and Mineral Resources of Curry County Oregon - Originally published in 1916, this important publication on Oregon Mining has not been available for nearly a century. Included are rare insights into the history, production and locations of dozens of gold mines in Curry County, Oregon, as well as detailed information on important Oregon mining districts in that area such as those at Agness, Bald Face Creek, Mule Creek, Boulder Creek, China Diggings, Collier Creek, Elk River, Gold Beach, Rock Creek, Sixes River and elsewhere. Particular attention is especially paid to the famous beach gold deposits of this portion of the Oregon Coast. **8.5" X 11", 140 ppgs. Retail Price: $11.99**

Chromite Mining in South West Oregon - Originally published in 1961, this important publication on Oregon Mining has not been available for nearly a century. Included are rare insights into the history, production and locations of nearly 300 chromite mines in South Western Oregon. **8.5" X 11", 184 ppgs. Retail Price: $14.99**

Mineral Resources of Douglas County Oregon - Originally published in 1972, this important publication on Oregon Mining has not been available for nearly forty years. Included are rare insights into the geology, history, production and locations of numerous gold mines and other mining properties in Douglas County, Oregon. **8.5" X 11", 124 ppgs. Retail Price: $11.99**

Mineral Resources of Coos County Oregon - Originally published in 1972, this important publication on Oregon Mining has not been available for nearly forty years. Included are rare insights into the geology, history, production and locations of numerous gold mines and other mining properties in Coos County, Oregon. **8.5" X 11", 100 ppgs. Retail Price: $11.99**

Mineral Resources of Lane County Oregon - Originally published in 1938, this important publication on Oregon Mining has not been available for nearly seventy five years. Included are extremely rare insights into the geology and mines of Lane County, Oregon, in particular in the Bohemia, Blue River, Oakridge, Black Butte and Winberry Mining Districts. **8.5" X 11", 82 ppgs. Retail Price: $9.99**

Mineral Resources of the Upper Chetco River of Oregon: Including the Kalmiopsis Wilderness - Originally published in 1975, this important publication on Oregon Mining has not been available for nearly forty years. Withdrawn under the 1872 Mining Act since 1984, real insight into the minerals resources and mines of the Upper Chetco River has long been unavailable due to the remoteness of the area. Despite this, the decades of battle between property owners and environmental extremists over the last private mining inholding in the area has continued to pique the interest of those interested in mining and other forms of natural resource use. Gold mining began in the area in the 1850's and has a rich history in this geographic area, even if the facts surrounding it are little known. Included are twenty two rare photographs, as well as insights into the Becca and Morning Mine, the Emmly Mine (also known as Emily Camp), the Frazier Mine, the Golden Dream or Higgins Mine, Hustis Mine, Peck Mine and others. **8.5" X 11", 64 ppgs. Retail Price: $8.99**

Gold Dredging in Oregon - Originally published in 1939, this important publication on Oregon Mining has not been available for nearly seventy five years. Included are extremely rare insights into the history and day to day operations of the dragline and bucketline gold dredges that once worked the placer gold fields of South West and North East Oregon in decades gone by. Also included are details into the areas that were worked by gold dredges in Josephine, Jackson, Baker and Grant counties, as well as the economic factors that impacted this mining method. This volume also offers a unique look into the values of river bottom land in relation to both farming and mining, in how farm lands were mined, re-soiled and reclamated after the dredges worked them. Featured are hard to find maps of the gold dredge fields, as well as rare photographs from a bygone era. **8.5" X 11", 86 ppgs. Retail Price: $8.99**

Quick Silver Mining in Oregon - Originally published in 1963, this important publication on Oregon Mining has not been available for over fifty years. This publication includes details into the history and production of Elemental Mercury or Quicksilver in the State of Oregon. **8.5" X 11", 238 ppgs. Retail Price: $15.99**

Mines of the Greenhorn Mining District of Grant County Oregon - Originally published in 1948, this important publication on Oregon Mining has not been available for over sixty five years. In this publication are rare insights into the mines of the famous Greenhorn Mining District of Grant County, Oregon, especially the famous Morning Mine. Also included are details on the Tempest, Tiger, Bi-Metallic, Windsor, Psyche, Big Johnny, Snow Creek, Banzette and Paramount Mines, as well as prospects in the vicinities in the famous mining areas of Mormon Basin, Vinegar Basin and Desolation Creek. Included are hard to find mine maps and dozens of rare photographs from the bygone era of Grant County's rich mining history. **8.5" X 11", 72 ppgs. Retail Price: $9.99**

Geology of the Wallowa Mountains of Oregon: Part I (Volume 1) - Originally published in 1938, this important publication on Oregon Mining has not been available for nearly seventy five years. Included are details on the geology of this unique portion of North Eastern Oregon. This is the first part of a two book series on the area. Accompanying the text are rare photographs and historic maps. 8.5" X 11", 92 ppgs. **Retail Price: $9.99**

Geology of the Wallowa Mountains of Oregon: Part II (Volume 2) - Originally published in 1938, this important publication on Oregon Mining has not been available for nearly seventy five years. Included are details on the geology of this unique portion of North Eastern Oregon. This is the first part of a two book series on the area. Accompanying the text are rare photographs and historic maps. 8.5" X 11", 94 ppgs. **Retail Price: $9.99**

Field Identification of Minerals For Oregon Prospectors - Originally published in 1940, this important publication on Oregon Mining has not been available for nearly seventy five years. Included in this volume is an easy system for testing and identifying a wide range of minerals that might be found by prospectors, geologists and rockhounds in the State of Oregon, as well as in other locales. Topics include how to put together your own field testing kit and how to conduct rudimentary tests in the field. This volume is written in a clear and concise way to make it useful even for beginners. **8.5" X 11", 158 ppgs. Retail Price: $14.99**

The Bohemia Mining District of Oregon - Originally published in 1900, this important publication on Oregon Mining has not been available for over a century. Included in this volume are important insights into the famous Bohemia Mining District of Oregon, including the histories and locations of important gold mines in the area such as the Ophir Mine, Clarence, Acturas, Peek-a-boo, White Swan, Combination Mine, the Musick Mine, The California, White Ghost, The Mystery, Wall Street, Vesuvius, Story, Lizzie Bullock, Delta, Elsie Dora, Golden Slipper, Broadway, Champion Mine, Knott, Noonday, Helena, White Wings, Riverside and others. Also included are notes on the nearby Blue River Mining District. **8.5" X 11", 58 ppgs. Retail Price: $9.99**

The Gold Fields of Eastern Oregon - Unavailable since 1900, this publication was originally compiled by the Baker City Chamber of Commerce Offering important insights into the gold mining history of Eastern Oregon, "The Gold Fields of Eastern Oregon" sheds a rare light on many of the gold mines that were operating at the turn of the 19th Century in Baker County and Grant County in North Eastern Oregon. Some of the areas featured include the Cable Cove District, Baisely-Elhorn, Granite, Red Boy, Bonanza, Susanville, Sparta, Virtue, Vaughn, Sumpter, Burnt River, Rye Valley and other mining districts. Included is basic information on not only many gold mines that are well known to those interested in Eastern Oregon mining history, but also many mines and prospects which have been mostly lost to the passage of time. Accompanying are numerous rare photos **8.5" X 11", 78 ppgs. Retail Price: $10.99**

Gold Mining in Eastern Oregon - Originally published in 1938, this important publication on Oregon Mining has not been available for over a century. Included in this volume are important insights into the famous mining districts of Eastern Oregon during the late 1930's. Particular attention is given to those gold mines with milling and concentrating facilities in the Greenhorn, Red Boy, Alamo, Bonanza, Granite, Cable Cove, Cracker Creek, Virtue, Keating, Medical Springs, Sanger, Sparta, Chicken Creek, Mormon Basin, Connor Creek, Cornucopia and the Bull Run Mining Districts. Some of the mines featured include the Ben Harrison, North Pole-Columbia, Highland Maxwell, Baisley-Elkhorn, White Swan, Balm Creek, Twin Baby, Gem of Sparta, New Deal, Gleason, Gifford-Johnson, Cornucopia, Record, Bull Run, Orion and others. Of particular interest are the mill flow sheets and descriptions of milling operations of these mines. **8.5" X 11", 68 ppgs. Retail Price: $8.99**

The Gold Belt of the Blue Mountains of Oregon - Originally published in 1901, this important publication on Oregon Mining has not been available for over a century. Included in this volume are rare insights into the gold deposits of the Blue Mountains of North East Oregon, including the history of their early discovery and early production. Extensive details are offered on this important mining area's mineralogy and economic geology, as well as insights into nearby gold placers, silver deposits and copper deposits. Featured are the Elkhorn and Rock Creek mining districts, the Pocahontas district, Auburn and Minersville districts, Sumpter and Cracker Creek, Cable Cove, the Camp Carson district, Granite, Alamo, Greenhorn, Robinsonville, the Upper Burnt River Valley and Bonanza districts, Susanville, Quartzburg, Canyon Creek, Virtue, the Copper Butte district, the North Powder River, Sparta, Eagle Creek, Cornucopia, Pine Creek, Lower Powder River, the Upper Snake River Canyon, Rye Valley, Lower Burnt River Valley, Mormon Basin, the Malheur and Clarks Creek districts, Sutton Creek and others. Of particular interest are important details on numerous gold mines and prospects in these mining districts, including their locations, histories, geology and other important information, as well as information on silver, copper and fire opal deposits. **8.5" X 11", 250 ppgs. Retail Price: $24.99**

Mining in the Cascades Range of Oregon - Originally published in 1938, this important publication on Oregon Mining has not been available for over seventy five years. Included in this volume are rare insights into the gold mines and other types of metal mines in the Cascades Mountain Range of Oregon. Some of the important mining areas covered include the famous Bohemia Mining District, the North Santiam Mining District, Quartzville Mining District, Blue River Mining District, Fall Creek Mining District, Oakridge District, Zinc District, Buzzard-Al Sarena District, Grand Cove, Climax District and Barron Mining District. Of particular interest are important details on over 100 mines and prospects in these mining districts, including their locations, histories, geology and other important information. **8.5" X 11", 170 ppgs. Retail Price: $14.99**

Beach Gold Placers of the Oregon Coast - Originally published in 1934, this important publication on Oregon Mining has not been available for over 80 years. Included in this volume are rare insights into the beach gold deposits of the State of Oregon, including their locations, occurance, composition and geology. Of particular interest is information on placer platinum in Oregon's rich beach deposits. Also included are the locations and other information on some famous Oregon beach mines, including the Pioneer, Eagle, Chickamin, Iowa and beach placer mines north of the mouth of the Rogue River. **8.5" X 11", 60 ppgs. Retail Price: $8.99**

Idaho Mining Books

Gold in Idaho - Unavailable since the 1940's, this publication was originally compiled by the Idaho Bureau of Mines and includes details on gold mining in Idaho. Included is not only raw data on gold production in Idaho, but also valuable insight into where gold may be found in Idaho, as well as practical information on the gold bearing rocks and other geological features that will assist those looking for placer and lode gold in the State of Idaho. This volume also includes thirteen gold maps that greatly enhance the practical usability of the information contained in this small book detailing where to find gold in Idaho. **8.5" X 11", 72 ppgs. Retail Price: $9.99**

Geology of the Couer D'Alene Mining District of Idaho - Unavailable since 1961, this publication was originally compiled by the Idaho Bureau of Mines and Geology and includes details on the mining of gold, silver and other minerals in the famous Coeur D'Alene Mining District in Northern Idaho. Included are details on the early history of the Coeur D'Alene Mining District, local tectonic settings, ore deposit features, information on the mineral belts of the Osburn Fault, as well as detailed information on the famous Bunker Hill Mine, the Dayrock Mine, Galena Mine, Lucky Friday Mine and the infamous Sunshine Mine. This volume also includes sixteen hard to find maps. **8.5" X 11", 70 ppgs. Retail Price: $9.99**

The Gold Camps and Silver Cities of Idaho - Originally published in 1963, this important publication on Idaho Mining has not been available for nearly fifty years. Included are rare insights into the history of Idaho's Gold Rush, as well as the mad craze for silver in the Idaho Panhandle. Documented in fine detail are the early mining excitements at Boise Basin, at South Boise, in the Owyhees, at Deadwood, Long Valley, Stanley Basin and Robinson Bar, at Atlanta, on the famous Boise River, Volcano, Little Smokey, Banner, Boise Ridge, Hailey, Leesburg, Lemhi, Pearl, at South Mountain, Shoup and Ulysses, Yellow Jacket and Loon Creek. The story follows with the appearance of Chinese miners at the new mining camps on the Snake River, Black Pine, Yankee Fork, Bay Horse, Clayton, Heath, Seven Devils, Gibbonsville, Vienna and Sawtooth City. Also included are special sections on the Idaho Lead and Silver mines of the late 1800's, as well as the mining discoveries of the early 1900's that paved the way for Idaho's modern mining and mineral industry. Lavishly illustrated with rare historic photos, this volume provides a one of a kind documentary into Idaho's mining history that is sure to be enjoyed by not only modern miners and prospectors who still scour the hills in search of nature's treasures, but also those enjoy history and tromping through overgrown ghost towns and long abandoned mining camps. **8.5" X 11", 186 ppgs. Retail Price: $14.99**

Ore Deposits and Mining in North Western Custer County Idaho - Unavailable since 1913, this important publication was originally published by the Us Department of the Interior and has been unavailable for a century. Included are fine details on the geology, geography, gold placers and gold and silver bearing quartz veins of the mining region of North West Custer County, Idaho. Of particular interest is a rare look at the mines and prospects of the region, including those such as the Ramshorn Mine, SkyLark, Riverview, Excelsior, Beardsley, Pacific, Hoosier, Silver Brick, Forest Rose and dozens of others in the Bay Horse Mining District. Also covered are the mines of the Yankee Fork District such as the Lucky Boy, Badger, Black, Enterprise, Charles Dickens, Morrison, Golden Sunbeam, Montana, Golden Gate and others, as well as those in the Loon Mining District. **8.5" X 11", 126 ppgs. Retail Price: $12.99**

Gold Rush To Idaho - Unavailable since 1963, this important publication was originally published by the Idaho Bureau of Mines and has been unavailable for 50 years. "Gold Rush To Idaho" revisits the earliest years of the discovery of gold in Idaho Territory and introduces us to the conditions that the pioneer gold seekers met when they blazed a trail through the wilderness of Idaho's mountains and discovered the precious yellow metal at Oro Fino and Pierce. Subsequent rushes followed at places like Elk City, Newsome, Clearwater Station, Florence, Warrens and elsewhere. Of particular interest is a rare look at the hardships that the first miners in Idaho met with during their day to day existences and their attempts to bring law and order to their mining camps. 8.5" X 11", 88 ppgs. **Retail Price: $9.99**

The Geology and Mines of Northern Idaho and North Western Montana - Unavailable since 1909, this important publication was originally published by the Us Department of the Interior and has been unavailable for a century. Included are fine details on the geology and geography of the mining regions of Northern Idaho and North Western Montana. Of particular interest is a rare look at the mines and prospects of the region, including those in the Pine Creek Mining District, Lake Pend Oreille district, Troy Mining District, Sylvanite District, Cabinet Mining District, Prospect Mining District and the Missoula Valley. Some of the mines featured include the Iron Mountain, Silver Butte, Snowshoe, Grouse Mountain Mine and others. 8.5" X 11", 142 ppgs. **Retail Price: $12.99**

Mining in the Alturas Quadrangle of Blaine County Idaho - Unavailable since 1922, this important publication was originally published by the Idaho Bureau of Mines and has been unavailable for ninety years. Topics include the geology, rock formations and the formation of ore deposits in this important mining area of Idaho. Of particular focus is information on the local geology, quartz veins and ore deposits of this portion of Idaho. Included are hard to find details, including the descriptions and locations of numerous gold and silver mines in the area including the Silver King, Pilgrim, Columbia, Lone Jack, Sunbeam, Pride of the West, Lucky Boy, Scotia, Atlanta, Beaver-Bidwell and others mines and prospects. 8.5" X 11", 56 ppgs. **Retail Price: $8.99**

Mining in Lemhi County Idaho - Originally published in 1913, this important book on Idaho Mining has not been available to miners for over a century. Included are rare insights into hundreds of gold, silver, copper and other mines in this famous Idaho mining area. Details include the locations, geology, history, production and other facts of the mines of this region, not only gold and silver hardrock mines, but also gold placer mines, lead-silver deposits, copper mines, cobalt-nickel deposits, tungsten and tin mines . It is lavishly illustrated with hard to find photos of the period and rare mining maps. Some of the vicinities featured include the Nicholia Mining District, Spring Mountain District, Texas District, Blue Wing District, Junction District, McDevitt District, Pratt Creek, Eldorado District, Kirtley Creek, Carmen Creek, Gibbonsville, Indian Creek, Mineral Hill District, Mackinaw, Eureka District, Blackbird District, YellowJacket District, Gravel Range District, Junction District, Parker Mountain and other mining districts. 8.5" X 11", 226 ppgs. **Retail Price: $19.99**

Utah Mining Books

Fluorite in Utah - Unavailable since 1954, this publication was originally compiled by the USGS, State of Utah and U.S. Atomic Energy Commission and details the mining of fluorspar, also known as fluorite in the State of Utah. Included are details on the geology and history of fluorspar (fluorite) mining in Utah, including details on where this unique gem mineral may be found in the State of Utah. 8.5" X 11", 60 ppgs. **Retail Price: $8.99**

California Mining Books

The Tertiary Gravels of the Sierra Nevada of California - Mining historian Kerby Jackson introduces us to a classic mining work by Waldemar Lindgren in this important re-issue of The Tertiary Gravels of the Sierra Nevada of California. Unavailable since 1911, this publication includes details on the gold bearing ancient river channels of the famous Sierra Nevada region of California. 8.5" X 11", 282 ppgs. **Retail Price: $19.99**

The Mother Lode Mining Region of California - Unavailable since 1900, this publication includes details on the gold mines of California's famous Mother Lode gold mining area. Included are details on the geology, history and important gold mines of the region, as well as insights into historic mining methods, mine timbering, mining machinery, mining bell signals and other details on how these mines operated. Also included are insights into the gold mines of the California Mother Lode that were in operation during the first sixty years of California's mining history. 8.5" X 11", 176 ppgs. **Retail Price: $14.99**

Lode Gold of the Klamath Mountains of Northern California and South West Oregon - Unavailable since 1971, this publication was originally compiled by Preston E. Hotz and includes details on the lode mining districts of Oregon and California's Klamath Mountains. Included are details on the geology, history and important lode mines of the French Gulch, Deadwood, Whiskeytown, Shasta, Redding, Muletown, South Fork, Old Diggings, Dog Creek (Delta), Bully Choop (Indian Creek), Harrison Gulch, Hayfork, Minersville, Trinity Center, Canyon Creek, East Fork, New River, Denny, Liberty (Black Bear), Cecilville, Callahan, Yreka, Fort Jones and Happy Camp mining districts in California, as well as the Ashland, Rogue River, Applegate, Illinois River, Takilma, Greenback, Galice, Silver Peak, Myrtle Creek and Mule Creek districts of South Western Oregon. Also included are insights into the mineralization and other characteristics of this important mining region. 8.5" X 11", 100 ppgs. **Retail Price: $10.99**

Mines and Mineral Resources of Shasta County, Siskiyou County, Trinity County: California - Unavailable since 1915, this publication was originally compiled by the California State Mining Bureau and includes details on the gold mines of this area of Northern California. Also included are insights into the mineralization and other characteristics of this important mining region, as well as the location of historic gold mines. 8.5" X 11", 204 ppgs. Retail Price: $19.99

Geology of the Yreka Quadrangle, Siskiyou County, California - Unavailable since 1977, this publication was originally compiled by Preston E. Hotz and includes details on the geology of the Yreka Quadrangle of Siskiyou County, California. Also included are insights into the mineralization and other characteristics of this important mining region. 8.5" X 11", 78 ppgs. Retail Price: $7.99

Mines of San Diego and Imperial Counties, California - Originally published in 1914, this important publication on California Mining has not been available for a century. This publication includes important information on the early gold mines of San Diego and Imperial County, which were some of the first gold fields mined in California by early Spanish and Mexican miners before the 49ers came on the scene. Included are not only details on early mining methods in the area, production statistics and geological information, but also the location of the early gold mines that helped make California "The Golden State". Also included are details on the mining of other minerals such as silver, lead, zinc, manganese, tungsten, vanadium, asbestos, barite, borax, cement, clay, dolomite, fluospar, gem stones, graphite, marble, salines, petroleum, stronium, talc and others. 8.5" X 11", 116 ppgs. Retail Price: $12.99

Mines of Sierra County, California - Unavailable since 1920, this publication was originally compiled by the California State Mining Bureau and includes details on the gold mines of Sierra County, California. Also included are insights into the mineralization and other characteristics of this important mining region, as well as the location of historic gold mines. 8.5" X 11", 156 ppgs. Retail Price: $19.99

Mines of Plumas County, California - Unavailable since 1918, this publication was originally compiled by the California State Mining Bureau and includes details on the gold mines of Plumas County, California. Also included are insights into the mineralization and other characteristics of this important mining region, as well as the location of historic gold mines. 8.5" X 11", 200 ppgs. Retail Price: $19.99

Mines of El Dorado, Placer, Sacramento and Yuba Counties, California - Originally published in 1917, this important publication on California Mining has not been available for nearly a century. This publication includes important information on the early gold mines of El Dorado County, Placer County, Sacramento County and Yuba County, which were some of the first gold fields mined by the Forty-Niners during the California Gold Rush. Included are not only details on early mining methods in the area, production statistics and geological information, but also the location of the early gold mines that helped make California "The Golden State". Also included are insights into the early mining of chrome, copper and other minerals in this important mining area. 8.5" X 11", 204 ppgs. Retail Price: $19.99

Mines of Los Angeles, Orange and Riverside Counties, California - Originally published in 1917, this important publication on California Mining has not been available for nearly a century. This publication includes important information on the early gold mines of Los Angeles County, Orange County and Riverside County, which were some of the first gold fields mined in California by early Spanish and Mexican miners before the 49ers came on the scene. Included are not only details on early mining methods in the area, production statistics and geological information, but also the location of the early gold mines that helped make California "The Golden State". 8.5" X 11", 146 ppgs. Retail Price: $12.99

Mines of San Bernadino and Tulare Counties, California - Originally published in 1917, this important publication on California Mining has not been available for nearly a century. This publication includes important information on the early gold mines of San Bernadino and Tulare County, which were some of the first gold fields mined in California by early Spanish and Mexican miners before the 49ers came on the scene. Included are not only details on early mining methods in the area, production statistics and geological information, but also the location of the early gold mines that helped make California "The Golden State". Also included are details on the mining of other minerals such as copper, iron, lead, zinc, manganese, tungsten, vanadium, asbestos, barite, borax, cement, clay, dolomite, fluospar, gem stones, graphite, marble, salines, petroleum, stronium, talc and others. 8.5" X 11", 200 ppgs. Retail Price: $19.99

Chromite Mining in The Klamath Mountains of California and Oregon - Unavailable since 1919, this publication was originally compiled by J.S. Diller of the United States Department of Geological Survey and includes details on the chromite mines of this area of Northern California and Southern Oregon. Also included are insights into the mineralization and other characteristics of this important mining region, as well as the location of historic mines. Also included are insights into chromite mining in Eastern Oregon and Montana. 8.5" X 11", 98 ppgs. Retail Price: $9.99

Mines and Mining in Amador, Calaveras and Tuolumne Counties, California - Unavailable since 1915, this publication was originally compiled by William Tucker and includes details on the mines and mineral resources of this important California mining area. Included are details on the geology, history and important gold mines of the region, as well as insights into other local mineral resources such as asbestos, clay, copper, talc, limestone and others. Also included are insights into the mineralization and other characteristics of this important portion of California's Mother Lode mining region. 8.5" X 11", 198 ppgs. Retail Price: $14.99

The Cerro Gordo Mining District of Inyo County California - Unavailable since 1963, this publication was originally compiled by the United States Department of Interior. Included are insights into the mineralization and other characteristics of this important mining region of Southern California. Topics include the mining of gold and silver in this important mining district in Inyo County, California, including details on the history, production and locations of the Cerro Gordo Mine, the Morning Star Mine, Estelle Tunnel, Charles Lease Tunnel, Ignacio, Hart, Crosscut Tunnel, Sunset, Upper Newtown, Newtown, Ella, Perseverance, Newsboy, Belmont and other silver and gold mines in the Cerro Gordo Mining District. This volume also includes important insights into the fossil record, geologic formations, faults and other aspects of economic geology in this California mining district. 8.5" X 11", 104 ppgs. Retail Price: $10.99

Mining in Butte, Lassen, Modoc, Sutter and Tehama Counties of California - Unavailable since 1917, this publication was originally compiled by the United States Department of Interior. Included are insights into the mineralization and other characteristics of this important mining region of California. Topics include the mining of asbestos, chromite, gold, diamonds and manganese in Butte County, the mining of gold and copper in the Hayden Hill and Diamond Mountain mining districts of Lassen County, the mining of coal, salt, copper and gold in the High Grade and Winters mining districts of Modoc County, gold mining in Sutter County and the mining of gold, chromite, manganese and copper in Tehama County. This volume also includes the production records and locations of numerous mines in this important mining region. 8.5" X 11", 114 ppgs. Retail Price: $11.99

Mines of Trinity County California - Originally published in 1965, this important publication on California Mining has not been available for nearly fifty years. This publication includes important information on mines and mining in Trinity County, California, as well insights into the mineralization and geology of this important mining area in Northern California. Included are extensive details on hardrock and placer gold mines and prospects, including charts showing the locations of these historic mines.. 8.5" X 11", 144 ppgs. Retail Price: $12.99

Mines of Kern County California - Originally published in 1962, this important publication on California Mining has not been available for nearly fifty years. This publication includes important information on mines and mining in Kern County, California, as well insights into the mineralization and geology of this important mining area in California. Included are extensive details on hardrock and placer gold mines and prospects, including charts showing the locations of these historic mines. 8.5" X 11", 398 ppgs. Retail Price: $24.99

Mines of Calaveras County California - Originally published in 1962, this important publication on California Mining has not been available for nearly fifty years. This publication includes important information on mines and mining in Calaveras County, California, as well insights into the mineralization and geology of this important mining area in Northern California. Included are extensive details on hardrock and placer gold mines and prospects, including charts showing the locations of these historic mines. 8.5" X 11", 236 ppgs. Retail Price: $19.99

Lode Gold Mining in Grass Valley California - Unavailable since 1940, this publication was originally compiled by the United States Department of Interior. Included are insights into the gold mineralization and other characteristics of this important mining region of Nevada County, California. This volume also includes important insights into the geologic formations, faults and other aspects of economic geology in this California mining district. Of particular interest are the fine details on many hardrock gold mines in the area, including their locations, histories, development and mineralization. Some of the mines featured include the Gold Hill Mine, Massachusetts Hill, Boundary, Peabody, Golden Center, North Star, Omaha, Lone Jack, Homeward Bound, Hartery, Wisconsin, Allison Ranch, Phoenix, Kate Hayes, W.Y.O.D., Empire, Rich Hill, Daisy Hill, Orleans, Sultana, Centennial, Conlin, Ben Franklin, Crown Point and many others. 8.5" X 11", 148 ppgs. Retail Price: $12.99

Lode Mining in the Alleghany District of Sierra County California - Unavailable since 1913, this publication was originally compiled by the United States Department of Interior. Included are insights into the mineralization and other characteristics of this important mining region of Sierra County. Included are details on the history, production and locations of numerous hardrock gold mines in this famous California area, including the Tightner Mine, Minnie D., Osceola, Eldorado, Twenty One, Sherman, Kenton, Oriental, Rainbow, Plumbago, Irelan, Gold Canyon, North Fork, Federal, Kate Hardy and others. This volume also includes important insights into the fossil record, geologic formations, faults and other aspects of economic geology in this California mining district. 8.5" X 11", 48 ppgs. Retail Price: $7.99

Six Months In The Gold Mines During The California Gold Rush - Unavailable since 1850, this important work is a first hand account of one "49'ers" personal experience during the great California Gold Rush, shedding important light on one of the most exciting periods in the history of not only California, but also the world. Compiled from journals written between 1847 and 1849 by E. Gould Buffum, a native of New York, "Six Months In The Gold Mines During The California Gold Rush" offers a rare look into the day to day lives of the people who came to California to work in her gold mines when the state was still a great frontier. **8.5" X 11", 290 ppgs. Retail Price: $19.99**

Quartz Mines of the Grass Valley Mining District of California - Unavailable since 1867, this important publication has not been available since those days. This rare publication offers a short dissertation on the early hardrock mines in this important mining district in the California Mother Lode region between the 1850's and 1860's. Also included are hard to find details on the mineralization and locations of these mines, as well as how they were operated in those day. **8.5" X 11", 44 ppgs. Retail Price: $8.99**

Alaska Mining Books

Ore Deposits of the Willow Creek Mining District, Alaska - Unavailable since 1954, this hard to find publication includes valuable insights into the Willow Creek Mining District near Hatcher Pass in Alaska. The publication includes insights into the history, geology and locations of the well known mines in the area, including the Gold Cord, Independence, Fern, Mabel, Lonesome, Snowbird, Schroff-O'Neil, High Grade, Marion Twin, Thorpe, Webfoot, Kelly-Willow, Lane, Holland and others. **8.5" X 11", 96 ppgs. Retail Price: $9.99**

The Juneau Gold Belt of Alaska - Unavailable since 1906, this hard to find publication includes valuable insights into the gold mines around Juneau, Alaska. The publication includes important details into the history, geology and locations of the well known gold mines and prospects in the area, including those around Windham Bay, Holkham Bay, Port Snettisham, on Grindstone and Rhine Creeks, Gold Creek, Douglas Island, Salmon Creek, Lemon Creek, Nugget Creek, from the Mendenhall River to Berners Bay, McGinnis Creek, Montana Creek, Peterson Creek, Windfall Creek, the Eagle River, Yankee Basin, Yankee Curve, Kowee Creek and elsewhere. Not only are gold placer mines included, but also hardrock gold mines. **8.5" X 11", 224 ppgs. Retail Price: $19.99**

Arizona Mining Books

Mines and Mining in Northern Yuma County Arizona - Originally published in 1911, this important publication on Arizona Mining has not been available for over a hundred years. Included are rare insights into the gold, silver, copper and quicksilver mines of Yuma County, Arizona together with hard to find maps and photographs. Some of the mines and mining districts featured include the Planet Copper Mine, Mineral Hill, the Clara Consolidated Mine, Viati Mine, Copper Basin prospect, Bowman Mine, Quartz King, Billy Mack, Carnation, the Wardwell and Osbourne, Valensuella Copper, the Mariquita, Colonial Mine, the French American, the New York-Plomosa, Guadalupe, Lead Camp, Mudersbach Copper Camp, Yellow Bird, the Arizona Northern (Salome Strike), Bonanza (Harqua Hala), Golden Eagle, Hercules, Socorro and others. **8.5" X 11", 144 ppgs. Retail Price: $11.99**

The Aravaipa and Stanley Mining Districts of Graham County Arizona - Originally published in 1925, this important publication on Arizona Mining has not been available for nearly ninety years. Included are rare insights into the gold and silver mines of these two important mining districts, together with hard to find maps. **8.5" X 11", 140 ppgs. Retail Price: $11.99**

Gold in the Gold Basin and Lost Basin Mining Districts of Mohave County, Arizona - This volume contains rare insights into the geology and gold mineralization of the Gold Basin and Lost Basin Mining Districts of Mohave County, Arizona that will be of benefit to miners and prospectors. Also included is a significant body of information on the gold mines and prospects of this portion of Arizona. This volume is lavishly illustrated with rare photos and mining maps. **8.5" X 11", 188 ppgs. Retail Price: $19.99**

Mines of the Jerome and Bradshaw Mountains of Arizona - This important publication on Arizona Mining has not been available for ninety years. This volume contains rare insights into the geology and ore deposits of the Jerome and Bradshaw Mountains of Arizona that will be of benefit to miners and prospectors who work those areas. Included is a significant body of information on the mines and prospects of the Verde, Black Hills, Cherry Creek, Prescott, Walker, Groom Creek, Hassayampa, Bigbug, Turkey Creek, Agua Fria, Black Canyon, Peck, Tiger, Pine Grove, Bradshaw, Tintop, Humbug and Castle Creek Mining Districts. This volume is lavishly illustrated with rare photos and mining maps. **8.5" X 11", 218 ppgs. Retail Price: $19.99**

The Ajo Mining District of Pima County Arizona - This important publication on Arizona Mining has not been available for nearly seventy years. This volume contains rare insights into the geology and mineralization of the Ajo Mining District in Pima County, Arizona and in particular the famous New Cornelia Mine. **8.5" X 11", 126 ppgs. Retail Price: $11.99**

Mining in the Santa Rita and Patagonia Mountains of Arizona - Originally published in 1915, this important publication on Arizona Mining has not been available for nearly a century. Included are rare insights into hundreds of gold, silver, copper and other mines in this famous Arizona mining area. Details include the locations, geology, history, production and other facts of the mines of this region. **8.5" X 11", 394 ppgs. Retail Price: $24.99**

Mining in the Bisbee Quadrangle of Arizona - Originally published in 1906, this important publication on Arizona Mining has not been available for nearly a century. Included are rare insights into hundreds of gold, silver, copper and other mines in this famous Arizona mining area. Details include the locations, geology, history, production and other facts of the mines of this important mining region. **8.5" X 11", 188 ppgs. Retail Price: $14.99**

Montana Mining Books

A History of Butte Montana: The World's Greatest Mining Camp - First published in 1900 by H.C. Freeman, this important publication sheds a bright light on one of the most important mining areas in the history of The West. Together with his insights, as well as rare photographs of the periods, Harry Freeman describes Butte and its vicinity from its early beginnings, right up to its flush years when copper flowed from its mines like a river. At the time of publication, Butte, Montana was known worldwide as "The Richest Mining Spot On Earth" and produced not only vast amounts of copper, but also silver, gold and other metals from its mines. Freeman illustrates, with great detail, the most important mines in the vicinity of Butte, providing rare details on their owners, their history and most importantly, how the mines operated and how their treasures were extracted. Of particular interest are the dozens of rare photographs that depict mines such as the famous Anaconda, the Silver Bow, the Smoke House, Moose, Paulin, Buffalo, Little Minah, the Mountain Consolidated, West Greyrock, Cora, the Green Mountain, Diamond, Bell, Parnell, the Neversweat, Nipper, Original and many others. **8.5" X 11", 142 ppgs. Retail Price: $12.99**

The Butte Mining District of Montana - This important publication on Montana Mining has not been available for over a century. Included are rare insights into the gold, copper and silver mines of Butte, Montana together with hard to find maps and photographs. Some of the topics include the early history of gold, silver and copper mining in the Butte area, insight into the geology of its mining areas, the local distribution of gold, silver and copper ores, as well their composition and how to identify them. Also included are detailed facts about the mines in the Butte Mining District, including the famous Anaconda Mine, Gagnon, Parrot, Blue Vein, Moscow, Poulin, Stella, Buffalo, Green Mountain, Wake Up Jim, the Diamond-Bell Group, Mountain Consolidated, East Greyrock, West Greyrock, Snowball, Corra, Speculator, Adirondack, Miners Union, the Jessie-Edith May Group, Otisco, Iduna, Colorado, Lizzie, Cambers, Anderson, Hesperus, Preferencia and dozens of others. **8.5" X 11", 298 ppgs. Retail Price: $24.99**

Mines of the Helena Mining Region of Montana - This important publication on Montana Mining has not been available for over a century. Included are rare insights into the gold, copper and silver mines of the vicinity of Helena, Montana, including the Marysville Mining District, Elliston Mining District, Rimini Mining District, Helena Mining District, Clancy Mining District, Wickes Mining District, Boulder and Basin Mining Districts and the Elkhorn Mining District. Some of the topics include the early history of gold, silver and copper mining in the Helena area, insight into the geology of its mining areas, the local distribution of gold, silver and copper ores, as well their composition and how to identify them. Also included are detailed facts, history, geology and locations of over one hundred gold, silver and copper mines in the area . **8.5" X 11", 162 ppgs, Retail Price: $14.99**

Mines and Geology of the Garnet Range of Montana - This important publication on Montana Mining has not been available for over a century. Included are rare insights into the gold, copper and silver mines of the vicinity of this important mining area of Montana. Some of the topics include the early history of gold, silver and copper mining in the Garnet Mountains, insight into the geology of its mining areas, the local distribution of gold, silver and copper ores, as well their composition and how to identify them. Also included are detailed facts, history, geology and locations of numerous gold, silver and copper mines in the area . **8.5" X 11", 100 ppgs, Retail Price: $11.99**

Mines and Geology of the Philipsburg Quadrangle of Montana - This important publication on Montana Mining has not been available for over a century. Included are rare insights into the gold, copper and silver mines of the vicinity of this important mining area of Montana. Some of the topics include the early history of gold, silver and copper mining in the Philipsburg Quadrangle, insight into the geology of its mining areas, the local distribution of gold, silver and copper ores, as well their composition and how to identify them. Also included are detailed facts, history, geology and locations of over one hundred gold, silver and copper mines in the area **8.5" X 11", 290 ppgs, Retail Price: $24.99**

Geology of the Marysville Mining District of Montana - Included are rare insights into the mining geology of the Marysville Mining District. Some of the topics include the early history of gold, silver and copper mining in the area, insight into the geology of its mining areas, the local distribution of gold, silver and copper ores, as well their composition and how to identify them. Also included are detailed facts, history, geology and locations of gold, silver and copper mines in the area **8.5" X 11", 198 ppgs, Retail Price: $19.99**

<u>**The Geology and Mines of Northern Idaho and North Western Montana**</u>

See listing under Idaho.

Nevada Mining Books

<u>**The Bull Frog Mining District of Nevada**</u> - Unavailable since 1910, this publication was originally compiled by the United States Department of Interior. This volume also includes important insights into the geologic formations, faults and other aspects of economic geology in this Nevada mining district. Of particular interest are the fine details on many mines in the area, including their locations, histories, development and mineralization. Some of the mines featured include the National Bank Mine, Providence, Gibraltor, Tramps, Denver, Original Bullfrog, Gold Bar, Mayflower, Homestake-King and other mines and prospects. **8.5" X 11", 152 ppgs, Retail Price: $14.99**

<u>History of the Comstock Lode</u> - Unavailable since 1876, this publication was originally released by John Wiley & Sons. This volume also includes important insights into the famous Comstock Lode of Nevada that represented the first major silver discovery in the United States. During its spectacular run, the Comstock produced over 192 million ounces of silver and 8.2 million ounces of gold. Not only did the Comstock result in one of the largest mining rushes in history and yield immense fortunes for its owners, but it made important contributions to the development of the State of Nevada, as well as neighboring California. Included here are important details on not only the early development and history of the Comstock, but also rare early insight into its mines, ore and its geology.**8.5" X 11", 244 ppgs, Retail Price: $19.99**

Colorado Mining Books

<u>**Ores of The Leadville Mining District**</u> - Unavailable since 1926, this publication was originally compiled by the United States Department of Interior. This volume also includes important insights into the ores and mineralization of the Leadville Mining District in Colorado. Topics include historic ore prospecting methods, local geology, insights into ore veins and stockworks, the local trend and distribution of ore channels, reverse faults, shattered rock above replacement ore bodies, mineral enrichment in oxidized and sulphide zones and more. **8.5" X 11", 66 ppgs, Retail Price: $8.99**

<u>**Mining in Colorado**</u> - Unavailable since 1926, this publication was originally compiled by the United States Department of Interior. This volume also includes important insights into the mining history of Colorado from its early beginnings in the 1850's right up to the mid 1920's. Not only is Colorado's gold mining heritage included, but also its silver, copper, lead and zinc mining industry. Each mining area is treated separately, detailing the development of Colorado's mines on a county by county basis. **8.5" X 11", 284 ppgs, Retail Price: $19.99**

<u>Gold Mining in Gilpin County Colorado</u> - Unavailable since 1876, this publication was originally compiled by the Register Steam Printing House of Central City, Colorado. A rare glimpse at the gold mining history and early mines of Gilpin County, Colorado from their first discovery in the 1850's up to the "flush years" of the mid 1870's. Of particular interest is the history of the discovery of gold in Gilpin County and details about the men who made those first strikes. Special focus is given to the early gold mines and first mining districts of the area, many of which are not detailed in other books on Colorado's gold mining history. **8.5" X 11", 156 ppgs, Retail Price: $12.99**

<u>Mining in the Gold Brick Mining District of Colorado</u> - Important insights into the history of the Gold Brick Mining District, as well as its local geography and economic geology. Also included are the histories and locations of historic mines in this important Colorado Mining District, including the Cortland, Carter, Raymond, Gold Links, Sacramento, Bassick, Sandy Hook, Chronicle, Grand Prize, Chloride, Granite Mountain, Lucille, Gray Mountain, Hilltop, Maggie Mitchell, Silver Islet, Revenue, Roosevelt, Carbonate King and others. In addition to hardrock mining, are also included are details on gold placer mining in this portion of Colorado. **8.5" X 11", 140 ppgs, Retail Price: $12.99**

Washington Mining Books

<u>**The Republic Mining District of Washington**</u> - Unavailable since 1910, this important publication was originally published by the Washington Geologic Survey and has been unavailable for a century. Topics include the geology, rock formations and the formation of ore deposits in this important mining area of Washington State. Also included are hard to find details on the geology, history and locations of dozens of mines in the area. Some of the mines featured include the New Republic Mine, Ben Hur, Morning Glory, the South Republic Mine, Quilp, Surprise, Black Tail, Lone Pine, San Poil, Mountain Lion, Tom Thumb, Elcaliph and many others. **8.5" X 11", 94 ppgs, Retail Price: $10.99**

The Myers Creek and Nighthawk Mining Districts of Washington - Unavailable since 1911, this important publication was originally published by the Washington Geologic Survey and has been unavailable for a century. Topics include the geology, rock formations and the formation of ore deposits in these important mining areas of Washington State. Also included are hard to find details on the geology, history and locations of dozens of mines in the area. Some of the mines featured include the Grant Mine, Monterey, Nip and Tuck, Myers Creek, Number Nine, Neutral, Rainbow, Aztec, Crystal Butte, Apex, Butcher Boy, Molson, Mad River, Olentangy, Delate, Kelsey, Golden Chariot, Okanogan, Ohio, Forty-Ninth Parallel, Nighthawk, Favorite, Little Chopaka, Summit, Number One, California, Peerless, Caaba, Prize Group, Ruby, Mountain Sheep, Golden Zone, Rich Bar, Similkameen, Kimberly, Triune, Hiawatha, Trinity, Hornsilver, Maquae, Bellevue, Bullfrog, Palmer Lake, Ivanhoe, Copper World and many others.
8.5″ X 11″, 136 ppgs, Retail Price: $12.99

The Blewett Mining District of Washington - Unavailable since 1911, this important publication was originally published by the Washington Geologic Survey and has been unavailable for a century. Topics include the geology, rock formations and the formation of ore deposits in this important mining area of Washington State. Also included are hard to find details on the geology, history and locations of dozens of mines in the area. Some of the mines featured include the Washington Meteor, Alta Vista, Pole Pick, Blinn, North Star, Golden Eagle, Tip Top, Wilder, Golden Guinea, Lucky Queen, Blue Bell, Prospect, Homestake, Lone Rock, Johnson, and others. **8.5″ X 11″, 134 ppgs, Retail Price: $12.99**

Silver Mining In Washington - Unavailable since 1955, this important publication was originally published by the Washington Geologic Survey. Featured are the hard to find locations and details pertaining to Washington's silver mines. **8.5″ X 11″, 180 ppgs, Retail Price: $15.99**

The Mines of Snohomish County Washington - Unavailable since 1942, this important publication was originally published by the Washington Geologic Survey and has been unavailable for seventy years. Featured are details on a large number of gold, silver, copper, lead and other metallic mineral mines. Included are the locations of each historic mine, along with information on the commodity produced.
8.5″ X 11″, 98 ppgs, Retail Price: $10.99

The Mines of Chelan County Washington - Unavailable since 1943, this important publication was originally published by the Washington Geologic Survey and has been unavailable for seventy years. Featured are details on a large number of gold, silver, copper, lead and other metallic mineral mines. Included are the locations of each historic mine, along with information on the commodity. **8.5″ X 11″, 88 ppgs, Retail Price: $9.99**

Metal Mines of Washington - Unavailable since 1921, this important publication was originally published by the Washington Geologic Survey and has been unavailable for nearly ninety years. Widely considered a masterpiece on the Washington Mining Industry, "Metal Mines of Washington" sheds light on the important details of Washington's early mining years. Featured are details on hundreds of gold, silver, copper, lead and other metallic mineral mines. Included are hard to find details on the mineral resources of this state, as well as the locations of historic mines. Lavishly illustrated with maps and historic photos and complete with a glossary to explain any technical terms found in the text, this is one of the most important works on mining in the State of Washington. No prospector or miner should be without it if they are interested in mining in Washington. **8.5″ X 11″, 396 ppgs, Retail Price: $24.99**

Gem Stones In Washington - Unavailable since 1949, this important publication was originally published by the Washington Geologic Survey and has been unavailable since first published. Included are details on where to find naturally occurring gem stones in the State of Washington, including quartz crystal, amethyst, smoky quartz, milky quartz, agates, bloodstone, carnelian, chert, flint, jasper, onyx, petrified wood, opal, fire opal, hyalite and others. **8.5″ X 11″, 54 ppgs, Retail Price: $8.99**

The Covada Mining District of Washington - Unavailable since 1913, this important publication was originally published by the Washington Geologic Survey and has been unavailable for a century. Topics include the geology, rock formations and the formation of ore deposits in this important mining area of Washington State. Also included are hard to find details on the geology, history and locations of dozens of mines in the area. Some of the mines featured include the Admiral, Advance, Algonkian, Big Bug, Big Chief, Big Joker, Black Hawk, Black Tail, Black Thorn, Captain, Cherokee Strip, Colorado, Dan Patch, Dead Shot, Etta, Good Ore, Greasy Run, Great Scott, Idora, IXL, Jay Bird, Kentucky Bell, King Solomon, Laurel, Laura S, Little Jay, Meteor, Neglected, Northern Light, Old Nell, Plymouth Rock, Polaris, Quandary, Reserve, Shoo Fly, Silver Plume, Three Pines, Vernie, White Rose and dozens of others. **8.5″ X 11″, 114 ppgs, Retail Price: $10.99**

The Index Mining District of Washington - Unavailable since 1912, this important publication was originally published by the Washington Geologic Survey and has been unavailable for a century. Topics include the geology, rock formations and the formation of ore deposits in this important mining area of Washington State. Also included are hard to find details on the geology, history and locations of dozens of mines in the area. Some of the mines featured include the Sunset, Non-Pareil, Ethel Consolidated, Kittaning, Merchant, Homestead, Co-operative, Lost Creek, Uncle Sam, Calumet, Florence-Rae, Bitter Creek, Index Peacock, Gunn Peak, Helena, North Star, Buckeye. Copper Bell, Red Cross and others. **8.5″ X 11″, 114 ppgs, Retail Price: $11.99**

Mining & Mineral Resources of Stevens County Washington - Unavailable since 1920, this important publication was originally published by the Washington Geologic Survey and has been unavailable for a century. Topics include the geology, rock formations and the formation of ore deposits in these important mining areas of Washington State. Also included are hard to find details on the geology, history and locations of hundreds of mines in the area. 8.5" X 11", 372 ppgs, Retail Price: $24.99

The Mines and Geology of the Loomis Quadrangle Okanogan County, Washington - Unavailable since 1972, this important publication was originally published by the Washington Geologic Survey and has been unavailable for a century. Topics include the geology, rock formations and the formation of ore deposits in this important mining area of Washington State. Also included are hard to find details on the geology, history and locations of dozens of gold, copper, silver and other mines in the area. 8.5" X 11", 150 ppgs, Retail Price: $12.99

The Conconully Mining District of Okanogan County Washington - Unavailable since 1973, this important publication was originally published by the Washington Geologic Survey and has been unavailable for a century. Topics include the geology, rock formations and the formation of ore deposits in this important mining area of Washington State, which also includes Salmon Creek, Blue Lake and Galena. Also included are hard to find details on the geology, mining history and locations of dozens of mines in the area. Some of the mines include Arlington, Fourth of July, Sonny Boy, First Thought, Last Chance, War Eagle-Peacock, Wheeler, Mohawk, Lone Star, Woo Loo Moo Loo, Keystone, Hughes, Plant-Callahan, Johnny Boy, Leuena, Gubser, John Arthur, Tough Nut, Homestake, Key and many others 8.5" X 11", 68 ppgs, **Retail Price: $8.99**

Wyoming Mining Books

Mining in the Laramie Basin of Wyoming - Unavailable since 1909, this publication was originally compiled by the United States Department of Interior. Also included are insights into the mineralization and other characteristics of this important mining region, especially in regards to coal, limestone, gypsum, bentonite clay, cement, sand, clay and copper. 8.5" X 11", 104 ppgs, Retail Price: $11.99

New Mexico Mining Books

The Mogollon Mining District of New Mexico - Unavailable since 1927, this important publication was originally published by the US Department of Interior and has been unavailable for 80 years. Topics include the geology, rock formations and the formation of ore deposits in this important mining area in New Mexico. Of particular focus is information on the history and production of the ore deposits in this area, their form and structure, vein filling, their paragenesis, origins and ore shoots, as well as oxidation and supergene enrichment. Also included are hard to find details, including the descriptions and locations of numerous gold, silver and other types of mines, including the Eureka, Pacific, South Alpine, Great Western, Enterprise, Buffalo, Mountain View, Floride, Gold Dust, Last Chance, Deadwood, Confidence, Maud S., Deep Down, Little Fanney, Trilby, Johnson, Alberta, Comet, Golden Eagle, Cooney, Queen, the Iron Crown, Eberle, Clifton, Andrew Jackson mine, Mascot and others. 8.5" X 11", 144 ppgs, Retail Price: $12.99

The Percha Mining District of Kingston New Mexico - Unavailable since 1883, this important publication was originally published by the Kingston Tribune and has been unavailable for over one hundred and thirty five years. Having been written during the earliest years of gold and silver mining in the Percha Mining District, unlike other books on the subject, this work offers the unique perspective of having actually been written while the early mining history of this area was still being made. In fact, the work was written so early in the development of this area that many of the notable mines in the Percha District were less than a few years old and were still being operated by their original discoverers with the same enthusiasm as when they were first located. Included are hard to find details on the very earliest gold and silver mines of this important mining district near Kingston in Sierra County, New Mexico. 8.5" X 11", 68 ppgs, Retail Price: $9.99

East Coast Mining Books

The Gold Fields of the Southern Appalachians - Unavailable since 1895, this important publication was originally published by the US Department of Interior and has been unavailable for nearly 120 years. Topics include the geology, rock formations and the formation of ore deposits in this important mining area of the American South. Of particular focus is information on the history and statistics of the ore deposits in this area, their form and structure and veins. Also included are details on the placer gold deposits of the region. The gold fields of the Georgian Belt, Carolinian Belt and the South Mountain Mining District of North Carolina are all treated in descriptive detail. Included are hard to find details, including the descriptions and locations of numerous gold mines in Georgia, North Carolina and elsewhere in the American South. Also included are details on the gold belts of the British Maritime Provinces and the Green Mountains. 8.5" X 11", 104 ppgs, Retail Price: $9.99

Gold Rush Tales Series

Millions in Siskiyou County Gold - In this first volume of the "Gold Rush Tales" series, leading mining historian and editor Kerby Jackson, introduces us to the story of how millions of dollars worth of gold was discovered in Siskiyou County during the California Gold Rush. Lavishly illustrated with photos from the 19th Century, this hard to find information was first published in 1897 and sheds important light onto the gold rush era in Siskiyou County, California and the experiences of the men who dug for the gold and actually found it. **8.5" X 11", 82 ppgs, Retail Price: $9.99**

The California Rand in the Days of '49 - In this second volume of the "Gold Rush Tales" series, leading mining historian and editor Kerby Jackson, introduces us to four tales from the California Gold Rush. Lavishly illustrated with photos from the 19th Century, this hard to find information was first published in 1890's and includes the stories of "California's Rand", details about Chinese miners, how one early miner named Baker struck it rich and also the story of Alphonzo Bowers, who invented the first hydraulic gold dredge. **8.5" X 11", 54 ppgs, Retail Price: $9.99**

More Mining Books

Prospecting and Developing A Small Mine - Topics covered include the classification of varying ores, how to take a proper ore sample, the proper reduction of ore samples, alluvial sampling, how to understand geology as it is applied to prospecting and mining, prospecting procedures, methods of ore treatment, the application of drilling and blasting in a small mine and other topics that the small scale miner will find of benefit. **8.5" X 11", 112 ppgs, Retail Price: $11.99**

Timbering For Small Underground Mines - Topics covered include the selection of caps and posts, the treatment of mine timbers, how to install mine timbers, repairing damaged timbers, use of drift supports, headboards, squeeze sets, ore chute construction, mine cribbing, square set timbering methods, the use of steel and concrete sets and other topics that the small underground miner will find of benefit. This volume also includes twenty eight illustrations depicting the proper construction of mine timbering and support systems that greatly enhance the practical usability of the information contained in this small book. **8.5" X 11", 88 ppgs. Retail Price: $10.99**

Timbering and Mining - A classic mining publication on Hard Rock Mining by W.H. Storms. Unavailable since 1909, this rare publication provides an in depth look at American methods of underground mine timbering and mining methods. Topics include the selection and preservation of mine timbers, drifting and drift sets, driving in running ground, structural steel in mine workings, timbering drifts in gravel mines, timbering methods for driving shafts, positioning drill holes in shafts, timbering stations at shafts, drainage, mining large ore bodies by means of open cuts or by the "Glory Hole" system, stoping out ore in flat or low lying veins, use of the "Caving System", stoping in swelling ground, how to stope out large ore bodies, Square Set timbering on the Comstock and its modifications by California miners, the construction of ore chutes, stoping ore bodies by use of the "Block System", how to work dangerous ground, information on the "Delprat System" of stoping without mine timbers, construction and use of headframes and much more. This volume provides a reference into not only practical methods of mining and timbering that may be employed in narrow vein mining by small miners today, but also rare insights into how mines were being worked at the turn of the 19th Century. **8.5" X 11", 288 ppgs. Retail Price: $24.99**

A Study of Ore Deposits For The Practical Miner - Mining historian Kerby Jackson introduces us to a classic mining publication on ore deposits by J.P. Wallace. First published in 1908, it has been unavailable for over a century. Included are important insights into the properties of minerals and their identification, on the occurrence and origin of gold, on gold alloys, insights into gold bearing sulfides such as pyrites and arsenopyrites, on gold bearing vanadium, gold and silver tellurides, lead and mercury tellurides, on silver ores, platinum and iridium, mercury ores, copper ores, lead ores, zinc ores, iron ores, chromium ores, manganese ores, nickel ores, tin ores, tungsten ores and others. Also included are facts regarding rock forming minerals, their composition and occurrences, on igneous, sedimentary, metamorphic and intrusive rocks, as well as how they are geologically disturbed by dikes, flows and faults, as well as the effects of these geologic actions and why they are important to the miner. Written specifically with the common miner and prospector in mind, the book will help to unlock the earth's hidden wealth for you and is written in a simple and concise language that anyone can understand. **8.5" X 11", 366 ppgs. Retail Price: $24.99**

Mine Drainage - Unavailable since 1896, this rare publication provides an in depth look at American methods of underground mine drainage and mining pump systems. This volume provides a reference into not only practical methods of mining drainage that may be employed in narrow vein mining by small miners today, but also rare insights into how mines were being worked at the turn of the 19th Century. **8.5" X 11", 218 ppgs. Retail Price: $24.99**

Fire Assaying Gold, Silver and Lead Ores - Unavailable since 1907, this important publication was originally published by the Mining and Scientific Press and was designed to introduce miners and prospectors of gold, silver and lead to the art of fire assaying. Topics include the fire assaying of ores and products containing gold, silver and lead; the sampling and preparation of ore for an assay; care of the assay office, assay furnaces; crucibles and scorifiers; assay balances; metallic ores; scorification assays; cupelling; parting' crucible assays, the roasting of ores and more. This classic provides a time honored method of assaying put forward in a clear, concise and easy to understand language that will make it a benefit to even beginners. **8.5″ X 11″, 96 ppgs. Retail Price: $11.99**

Methods of Mine Timbering - Originally published in 1896, this important publication on mining engineering has not been available for nearly a century. Included are rare insights into historical methods of timbering structural support that were used in underground metal mines during the California that still have a practical application for the small scale hardrock miner of today. **8.5″ X 11″, 94 ppgs. Retail Price: $10.99**

The Enrichment of Copper Sulfide Ores - First published in 1913, it has been unavailable for over a century. Topics include the definition and types of ore enrichment, the oxidation of copper ores, the precipitation of metallic sulfides. Also included are the results of dozens of lab experiments pertaining to the enrichment of sulfide ores that will be of interest to the practical hard rock mine operator in his efforts to release the metallic bounty from his mine's ore. **8.5″ X 11″, 92 ppgs. Retail Price: $9.99**

A Study of Magmatic Sulfide Ores - Unavailable since 1914, this rare publication provides an in depth look at magmatic sulfide ores. Some of the topics included are the definition and classification of magmatic ores, descriptions of some magmatic sulfide ore deposits known at the time of publication including copper and nickel bearing pyrrohitic ore bodies, chalcopyrite-bornite deposits, pyritic deposits, magnetite-ileminite deposits, chromite deposits and magmatic iron ore deposits. Also included are details on how to recognize these types of ore deposits while prospecting for valuable hardrock minerals. **8.5″ X 11″, 138 ppgs. Retail Price: $11.99**

The Cyanide Process of Gold Recovery - Unavailable since 1894 and released under the name "The Cyanide Process: Its Practical Application and Economical Results", this rare publication provides an in depth look at the early use of cyanide leaching for gold recovery from hardrock mine ores. This volume provides a reference into the early development and use of cyanide leaching to recover gold. **8.5″ X 11″, 162 ppgs. Retail Price: $14.99**

California Gold Milling Practices - Unavailable since 1895 and released under the name "California Gold Practices", this rare publication provides an in depth look at early methods of milling used to reduce gold ores in California during the late 19th century. This volume provides a reference into the early development and use of milling equipment during the earliest years of the California Gold Rush up to the age of the Industrial Revolution. Much of the information still applies today and will be of use to small scale miners engaging in hardrock mining. **8.5″ X 11″, 104 ppgs. Retail Price: $10.99**

www.ingramcontent.com/pod-product-compliance
Lightning Source LLC
Chambersburg PA
CBHW080816180526
45168CB00006B/2474